Holt Algebra 2

Chapter 14 Resource Book

HOLT, RINEHART AND WINSTON
A Harcourt Education Company
Orlando • Austin • New York • San Diego • London

Printed in the United States of America

ISBN 0-03-042824-6

5 6 7 8 9 862 09 08

Contents

Date ______________

Dear Family,

In Chapter 14 your child will continue to study trigonometry, specifically graphing trigonometric functions, using trigonometric identities, and solving trigonometric equations.

The graphs of the basic trigonometric functions are all **periodic functions** that repeat in regular intervals called **cycles.** The length of the cycle is called the **period** of the function. The **amplitude** of the function is half the distance between the maximum and minimum values of the function. Here are the basic shapes of the trigonometric functions:

Graphs of Trigonometric Functions

Sine	Cosine	Tangent
$y = \sin x$	$y = \cos x$	$y = \tan x$
period: 2π amplitude: 1	period: 2π amplitude: 1	period: π amplitude: undefined
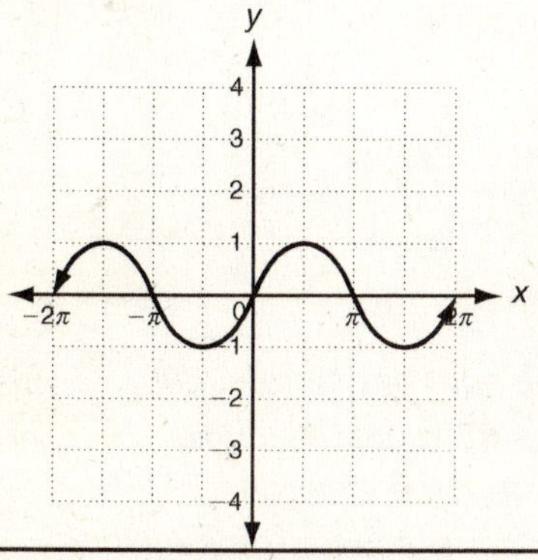	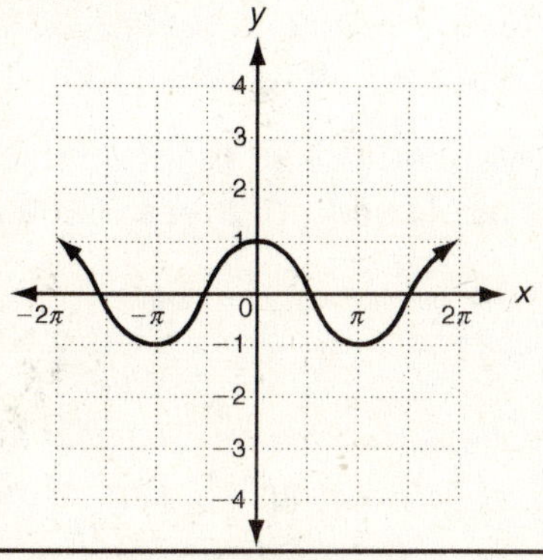	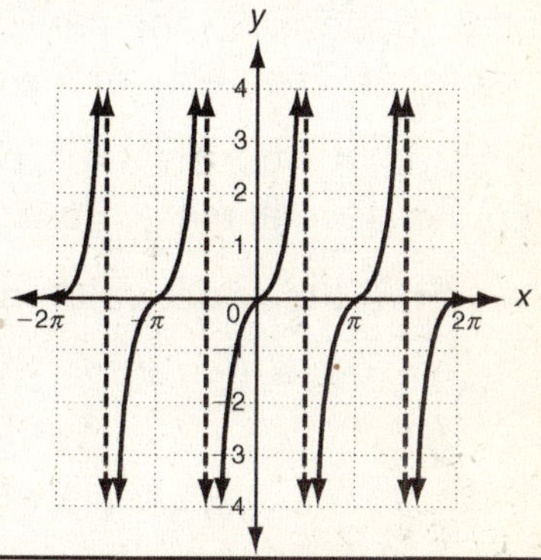

Cosecant	Secant	Cotangent
$y = \csc x$	$y = \sec x$	$y = \cot x$
period: 2π amplitude: undefined	period: 2π amplitude: undefined	period: π amplitude: undefined
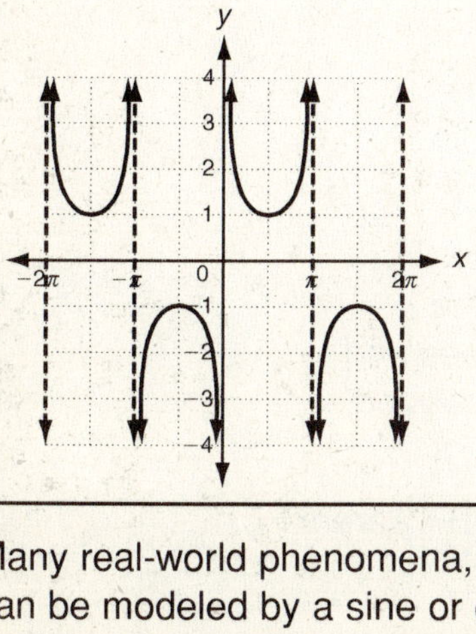	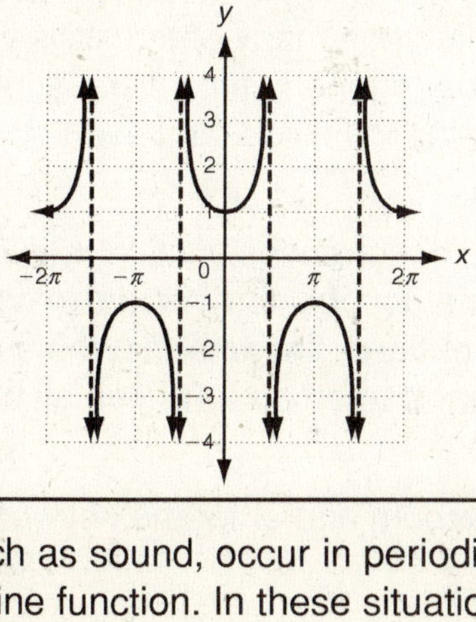	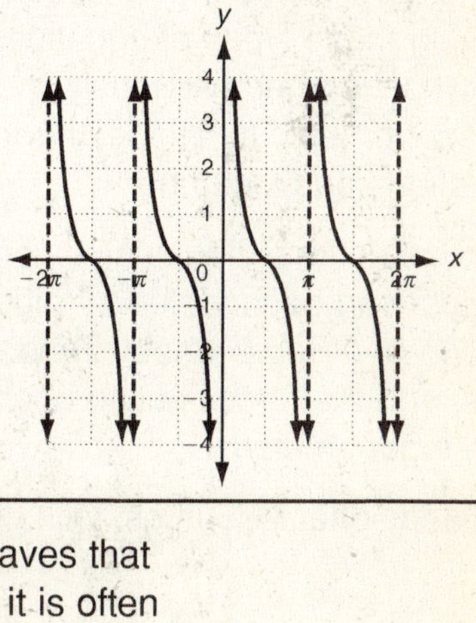

Many real-world phenomena, such as sound, occur in periodic waves that can be modeled by a sine or cosine function. In these situations, it is often useful to describe the number of cycles that occur in a given unit of time, called the **frequency** of the function.

As with all of the functions that your child has studied in Algebra 2, trigonometric functions can be transformed by stretching, compressing, and translating the graph. Vertical stretches and compressions will affect the amplitude of the function. Horizontal stretches and compressions affect the period and vertical asymptotes (if any). For trigonometric functions, a horizontal translation is also called a **phase shift.**

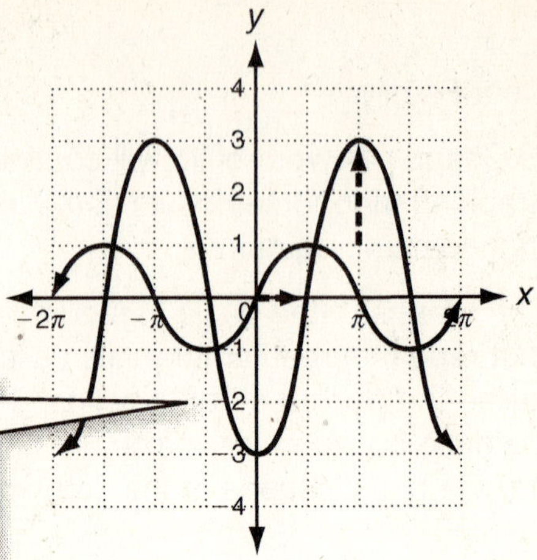

Compared to $y = \sin x$, the graph of $y = 3 \sin\left(x - \frac{\pi}{2}\right)$ is stretched vertically by a factor of 3 (changing the amplitude from 1 to 3), and is horizontally translated right by a phase shift of $\frac{\pi}{2}$.

An **identity** is a mathematical statement that is true for all values of the variable for which the statement is defined. For example, the statement $\cos^2 \theta + \sin^2 \theta = 1$ is true for any angle θ, and is therefore a trigonometric identity. (Specifically, the previous identity is one of several based on the Pythagorean Theorem, and hence called a Pythagorean Identity.)

In Chapter 14, your child will learn over two dozen trigonometric identities, and prove several more. Here's one application of these identities:

Find the exact value of sin 15°.

$\sin 15° = \sin(45° - 30°)$ — *Write 15° as a sum or difference of familiar angles from the unit circle.*

$= \sin 45° \cos 30° - \cos 45° \sin 30°$ — *One of the Difference Identities is sin (A − B) = sin A cos B − cos A sin B.*

$= \frac{\sqrt{2}}{2} \cdot \frac{\sqrt{3}}{2} - \frac{\sqrt{2}}{2} \cdot \frac{1}{2}$ — *Evaluate. (Refer to the unit circle in the Parent Letter for Chapter 13.)*

$= \frac{\sqrt{6}}{4} - \frac{\sqrt{2}}{4} = \frac{\sqrt{6} - \sqrt{2}}{4}$ — *Simplify.*

The chapter ends with a lesson on solving trigonometric equations. Solving these types of equations calls upon many of the skills your child has learned previously, including trigonometric inverses, graphs of trigonometric functions, trigonometric identities, and even quadratic equations and factoring.

Many situations that involve periodic occurrences, circular motion, or angles, can be modeled by trigonometric functions. Sound waves, the motion of a Ferris wheel, the depth of tides, the time that the sun rises and sets, and the forces acting upon a car parked on a hill can all be analyzed with trigonometry.

For additional resources, visit go.hrw.com and enter the keyword MB7 Parent.

Name ______________________ Date ____________ Class ____________

LESSON 14-1

Practice A

Graphs of Sine and Cosine

Identify whether each function is periodic. If the function is periodic, give the period.

1.

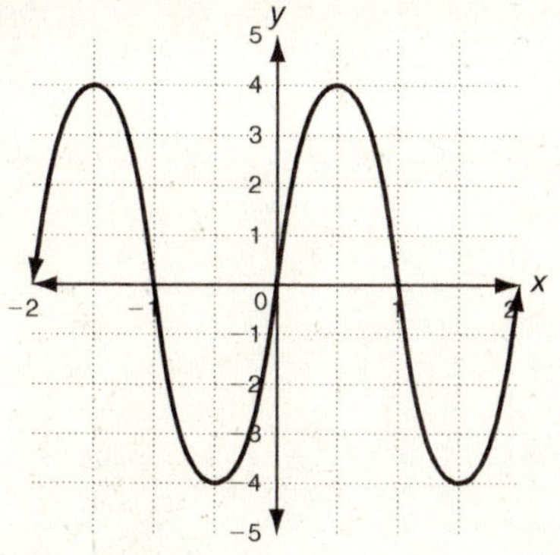

____

2.

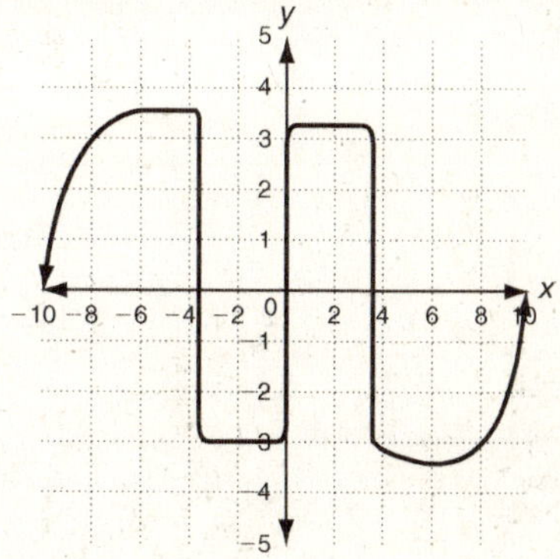

____

Use $f(x) = \sin x$ or $g(x) = \cos x$ as a guide. Identify the amplitude and period. Then graph each function.

3. $h(x) = -3\sin(4x)$

____

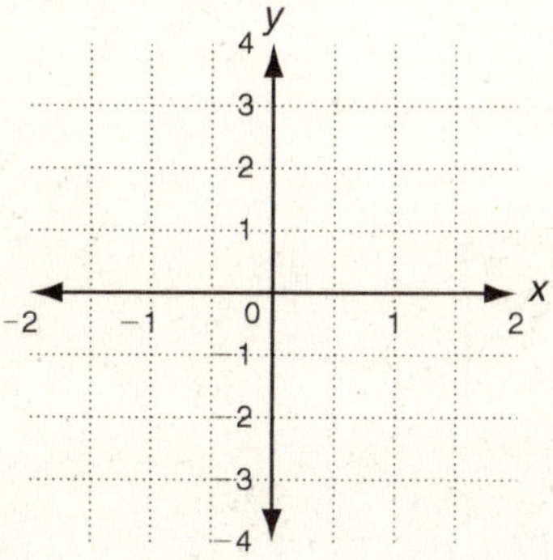

4. $p(x) = 2\cos(0.5x)$

____

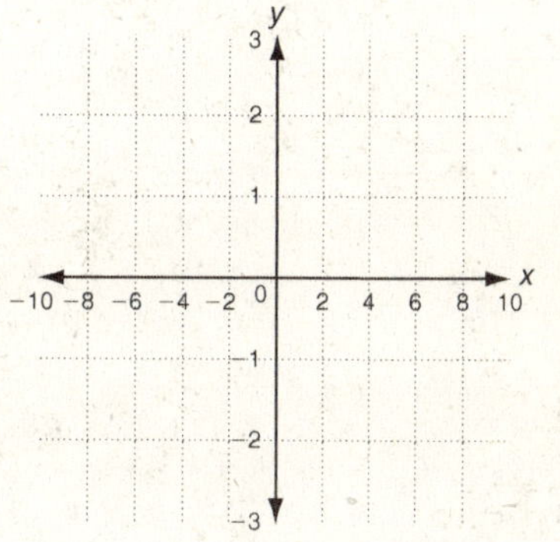

Use $f(x) = \cos x$ as a guide. Graph the function.

5. $g(x) = \cos(x - \pi)$

a. Identify the amplitude and period.

____

b. Identify the phase shift.

____

c. Identify the *x*-intercepts.

____

d. Identify the maximum and minimum values.

____

e. Use the information to graph the function.

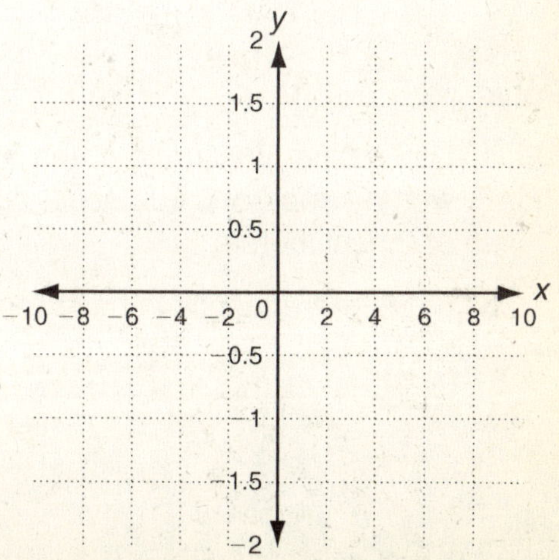

Name ______________________ Date __________ Class __________

LESSON 14-1

Practice B

Graphs of Sine and Cosine

Using $f(x) = \sin x$ or $g(x) = \cos x$ as a guide, graph each function. Identify the amplitude and period.

1. $b(x) = -5\sin \pi x$

____

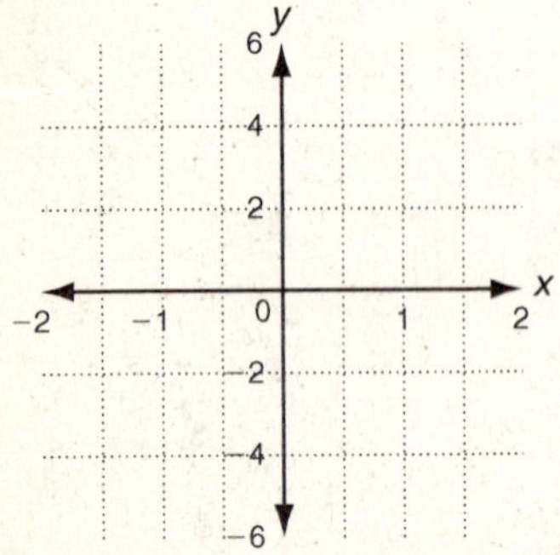

2. $k(x) = 3\cos 2\pi x$

____

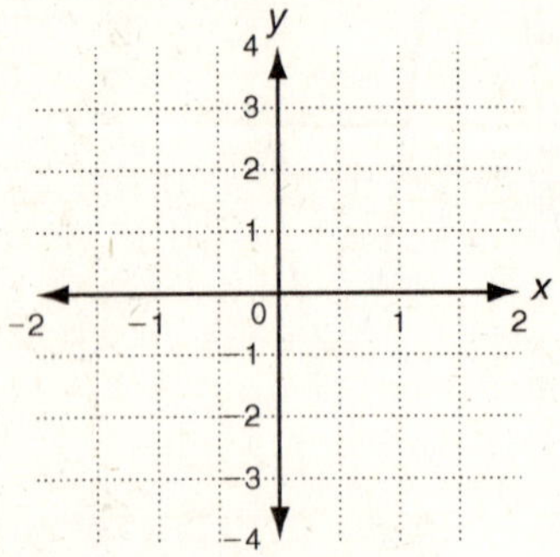

Using $f(x) = \sin x$ or $g(x) = \cos x$ as a guide, graph each function. Identify the *x*-intercepts and phase shift.

3. $h(x) = \sin\left(x + \frac{\pi}{4}\right)$

____

____

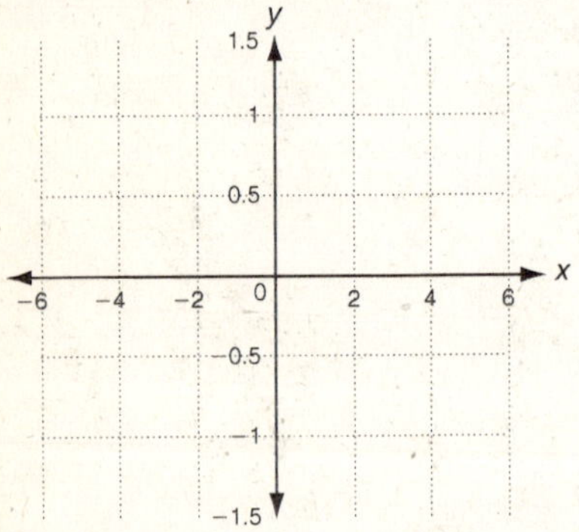

4. $h(x) = \cos\left(x - \frac{\pi}{4}\right)$

____

____

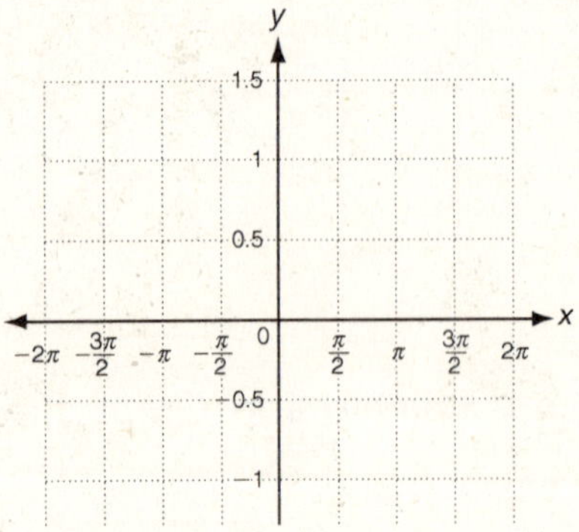

Solve.

5. a. Use a sine function to graph a sound wave with a period of 0.002 second and an amplitude of 2 centimeters.

b. Find the frequency in hertz for this sound wave.

____

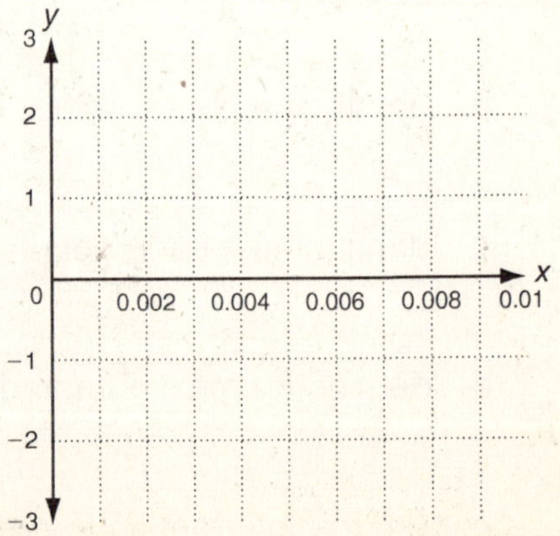

Name ______________________ Date __________ Class __________

LESSON **14-1**

Practice C
Graphs of Sine and Cosine

Using $f(x) = \sin x$ or $f(x) = \cos x$ as a guide, graph each function. Identify the amplitude, period, x-intercepts, and phase shift.

1. $h(x) = \frac{1}{2}\cos(-\pi x)$

____

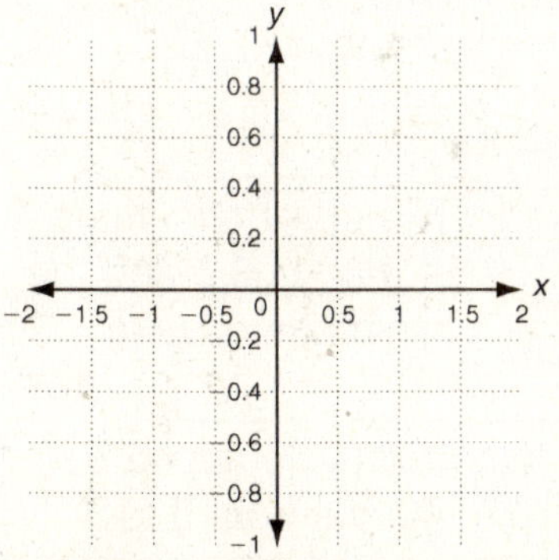

2. $q(x) = -\sin\left(\frac{\pi}{2}x\right)$

____

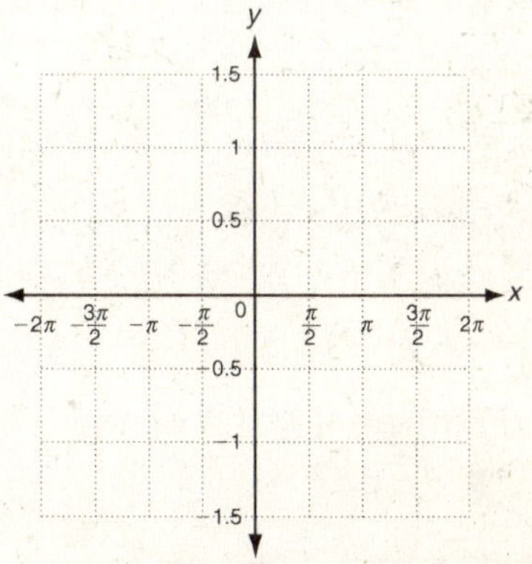

3. $c(x) = 3\cos(x + \pi)$

____

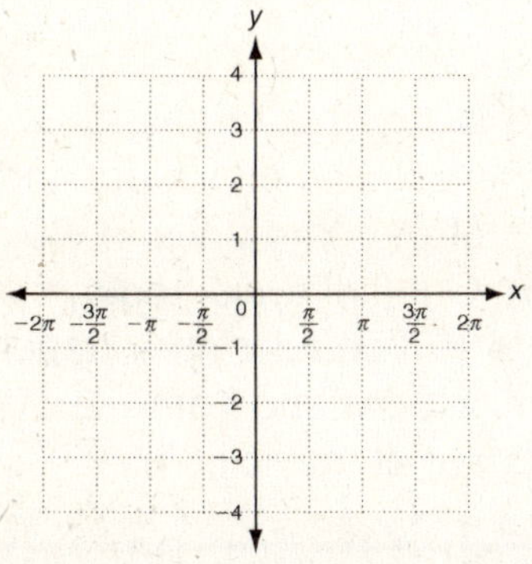

4. $h(x) = -2\sin\left(x - \frac{\pi}{2}\right)$

____

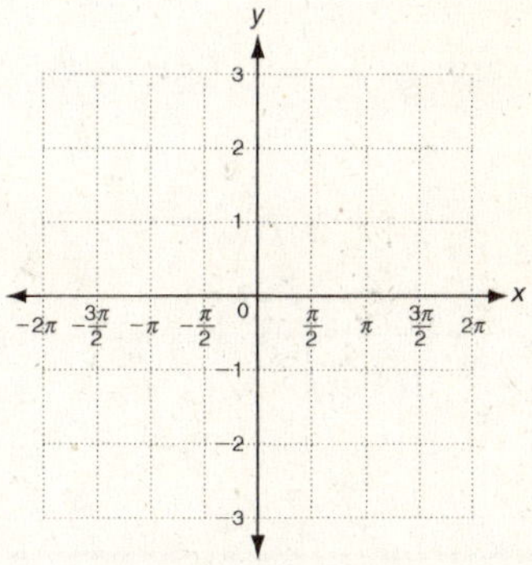

Solve.

5. A manual metronome is an inverted pendulum that helps musicians play to the beat. The number of centimeters, C, that the tip of the pendulum is from a tabletop can be modeled by $C(t) = 2\cos 4\pi t + 12$, where t is the time in seconds.

a. Graph the height of the pendulum tip for 2 periods.

b. How high is the pendulum when $t = \frac{1}{4}$ second?

____

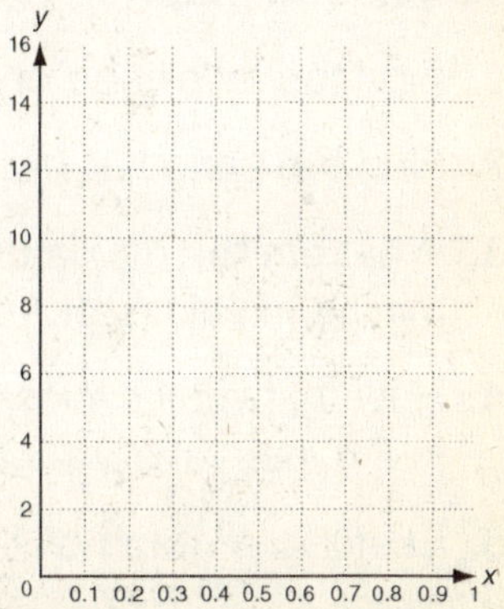

Name ______________________________ Date ___________ Class ___________

LESSON **14-1**

Reteach

Graphs of Sine and Cosine

Transformations of the sine and cosine functions change the amplitude and/or the period of the graph.

For $y = a\sin bx$ or $y = a\cos bx$:

- the amplitude is $|a|$,
- the period is $\frac{2\pi}{|b|}$.

> The amplitude is half the difference between the greatest and least values of the function.

> One full cycle appears in each period.

Use the graph of $f(x) = \sin x$ to sketch the graph of $g(x) = 0.5\sin 2x$.

Step 1 Compare $g(x) = 0.5\sin 2x$ to $y = a\sin bx$.
Find a to identify the amplitude.

$a = 0.5$ and $|0.5| = 0.5$, so the amplitude is 0.5.

> The maximum value of $g(x)$ is 0.5 and the minimum value is -0.5.

Step 2 Find b to identify the period.

$b = 2$, and $\frac{2\pi}{|b|} = \frac{2\pi}{|2|} = \pi$, so the period is π.

> One full cycle appears in the interval from 0 to π.

Step 3 Graph $f(x) = \sin x$.

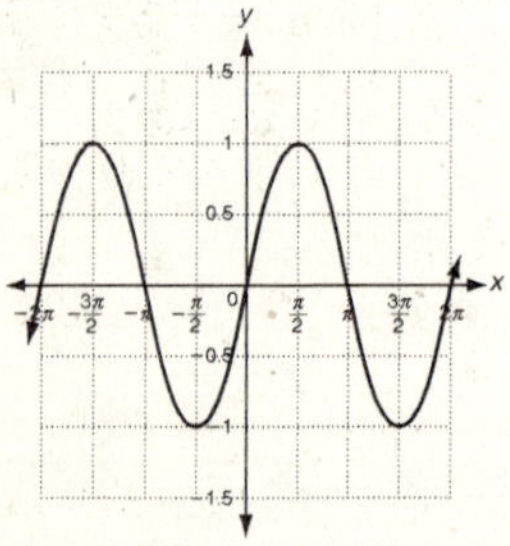

> The **amplitude** is 1. The maximum and minimum values of $f(x)$ are 1 and -1.
> The **period** is 2π. One full cycle appears in the interval from 0 to 2π. Two full cycles appear in the interval from -2π to 2π.
> The x-intercepts are at multiples of π.

Step 4 Graph $g(x) = 0.5\sin 2x$ on the same plane as $f(x)$.

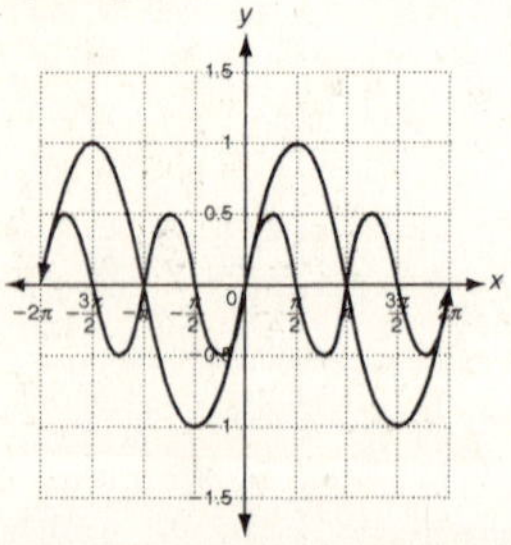

> The **amplitude** is 0.5. The maximum and minimum values of $g(x)$ are 0.5 and -0.5.
> The **period** is π. One full cycle appears in the interval from 0 to π. Two full cycles appear in the interval from 0 to 2π and from -2π to 0.
> The x-intercepts are at multiples of $\frac{\pi}{2}$.

Complete to graph $h(x) = 0.5\cos 2x$.

1. Find the amplitude of $h(x)$. $a =$ ____________________
2. Find the period of $h(x)$. $\frac{2\pi}{|b|} =$ ____________________
3. What are the maximum and minimum values of $h(x)$? ____________________
4. How many full cycles appear in the interval from 0 to π? ____________________
5. Sketch the graph of $f(x) = \cos x$. Then graph $h(x) = 0.5\cos 2x$.

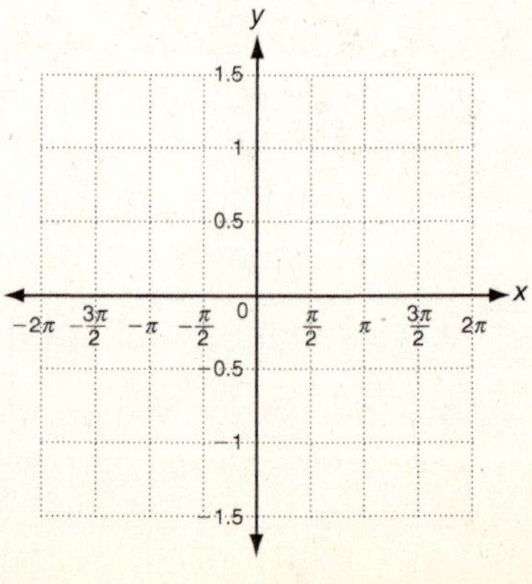

Name ______________________ Date ____________ Class ____________

LESSON 14-1

Reteach

Graphs of Sine and Cosine (continued)

A **phase shift** is a horizontal translation. Sine and cosine can be translated horizontally by $y = \sin(x - h)$ and $y = \cos(x - h)$.

A phase shift (or horizontal translation) of h units moves the graph left h units for $h < 0$ or right h units for $h > 0$.

Use the graph of $f(x) = \cos x$ to sketch the graph of $g(x) = \cos\left(x - \frac{\pi}{2}\right)$.

Step 1 Compare $g(x) = \cos\left(x - \frac{\pi}{2}\right)$ to $y = \mathbf{a} \cos \mathbf{b}x$.

Find the amplitude and period.
$a = 1$ and $|1| = 1$, so the amplitude is 1.
$b = 1$, and $\frac{2\pi}{|b|} = \frac{2\pi}{|1|} = 2\pi$, so the period is 2π.

> The amplitude and period of g are the same as for $y = \cos x$.

Step 2 Find h to identify the phase shift.
$x - h = x - \frac{\pi}{2}$, so $h = \frac{\pi}{2}$.
The phase shift is $\frac{\pi}{2}$ radians to the right.

> Because $h > 0$, the shift is to the right.

Step 3 Identify the first two positive x-intercepts.
The x-intercepts of $f(x) = \cos x$ occur at $\frac{\pi}{2}$ and $\frac{3\pi}{2}$.
The x-intercepts of $g(x)$ occur at $\frac{\pi}{2} + \frac{\pi}{2}$, or π, and $\frac{3\pi}{2} + \frac{\pi}{2}$, or 2π.

> Intercepts occur at integer multiples of π.

Step 4 Identify the maximum and minimum values.
The maxima and minima of $f(x) = \cos x$ occur at 0 and π.

The maxima and minima of $g(x)$ occur at $0 + \frac{\pi}{2}$, or $\frac{\pi}{2}$, and $\pi + \frac{\pi}{2}$, or $\frac{3\pi}{2}$.

Step 5 Graph $f(x) = \cos x$ and $g(x) = \cos\left(x - \frac{\pi}{2}\right)$.

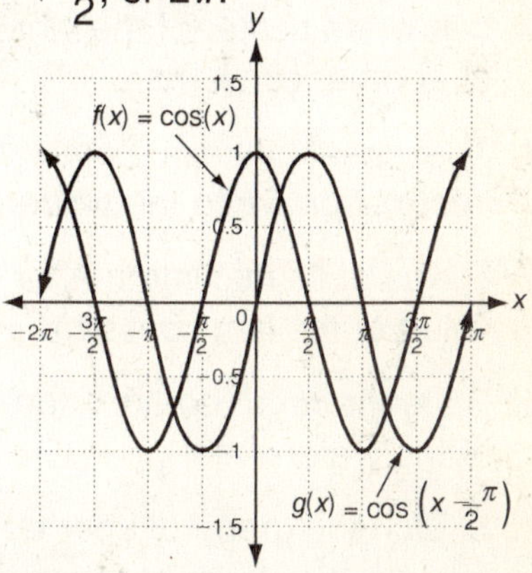

Use the graph of $f(x) = \sin x$ to sketch the graph of $g(x) = \sin\left(x - \frac{\pi}{2}\right)$.

6. Identify h. What is the phase shift?

____

7. Identify the x-intercepts from 0 to 2π.

____

8. Identify the maxima and minima from 0 to 2π.

____

9. Sketch the graphs of $f(x)$ and $g(x)$.

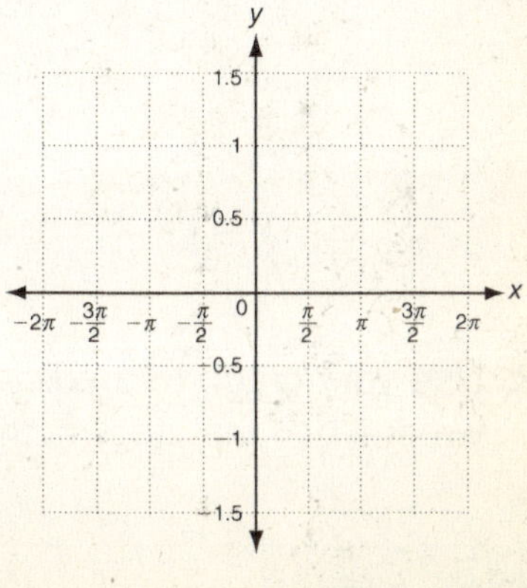

Name ______________________________ Date __________ Class __________

LESSON 14-1

Challenge

Sinusoidal Functions as Mathematical Models

In many real-world situations, a dependent variable repeats its values at regular intervals as the independent variable changes. In such cases, a sine or cosine function would be a reasonable mathematical model.

As represented at right, a waterwheel with a radius of 7 feet rotates at 6 revolutions per minute. You start a stopwatch when the point *P* on the rim of the wheel is at its lowest possible position. After 5 seconds, point *P* is at its greatest height.

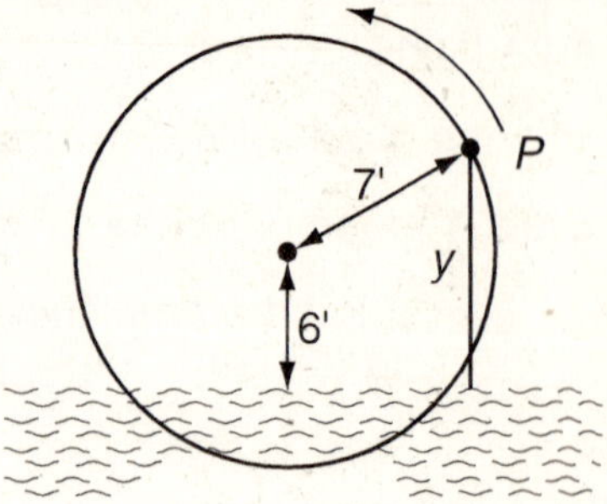

To write an equation that will model the distance, *y*, of point *P* from the surface of the water in terms of the number of seconds, *x*, on the stopwatch, assume that *y* varies sinusoidally with *x*, and begin with a sine equation.

1. In the general sine equation, $y = a \sin b(x - h) + k$, identify the constants that need to be determined. ______________

2. Since the center is 6 feet above the water, the horizontal line of symmetry of the sine curve will be 6 units above the *x*-axis. This information gives you the value of one of the constants. Write the value of this constant. ______________

3. Since the radius of the wheel is 7 feet, point *P* goes 7 feet above and below the center of the wheel. From this information, write the value of another constant. ______________

4. The wheel makes 6 complete revolutions every 60 seconds. From this information, write the value of another constant. ______________

5. *P* was at its maximum after 5 seconds. Use this information to write the value of the last constant. ______________

6. Write the required equation to model the movement of the waterwheel. ______________

7. Sketch the graph of your equation on the coordinate plane below.

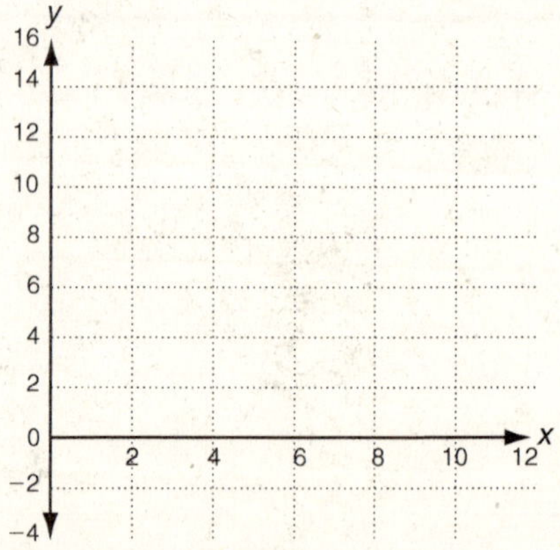

8. Use your graph to predict the height of point *P* above the surface of the water when the stopwatch shows 7 seconds.

____

Name ______________________ Date ____________ Class ____________

LESSON 14-1

Problem Solving

Graphs of Sine and Cosine

As a result of the tide, the depth of the water in the bay varies with time. According to the harbormaster, the depth can be modeled by the function $d(t) = 2.5 \cos\left(\frac{\pi}{6}\right)(t - 2) + 4$, where t is the number of hours since midnight, and d is the depth in meters.

1. Identify the following features of the graph of the function.

   a. Amplitude ____________

   b. Period ____________

   c. Phase shift ____________

   d. Vertical shift ____________

   e. What is the maximum depth and when does that occur?

   ____

   f. What is the minimum depth and when does that occur?

   ____

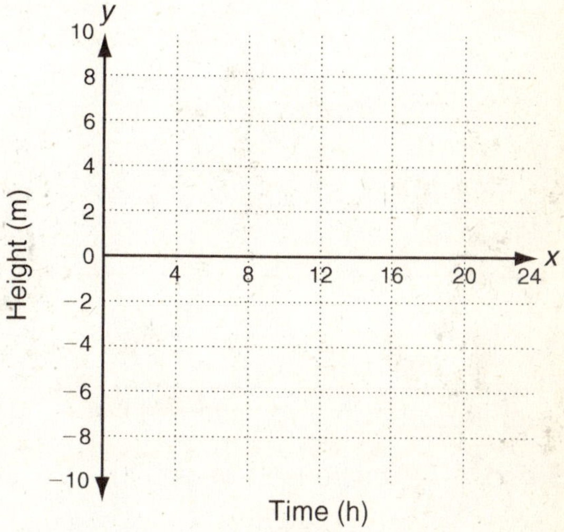

2. Graph the function for 24 hours.
3. How many high tides occur in 24 hours? ____________
4. Jim thinks that the amplitude and the vertical shift should be reversed.

   a. Write the equation for this new function. ____________

   b. Explain how water depth and frequency of tides would change using Jim's function.

   ____

   ____

5. Petra thinks that the original function is correct, except for the period, which should be 24. If so, how many hours after midnight would the first high and low tides occur? ____________

Jim looks at equations for the depth of water in different locations. Choose the letter for the best answer.

6. Which function represents a location that will dry out?

   A $d(t) = 5 \cos \frac{\pi}{12}(t - 1) + 4$

   B $d(t) = 4 \cos \frac{\pi}{8}(t - 2) + 4.5$

   C $d(t) = 3 \cos \frac{\pi}{6}(t - 3) + 5$

   D $d(t) = 2 \cos \frac{\pi}{4}(t - 4) + 5.5$

7. How deep is the water at 1:00 A.M. in a location where the water depth t hours after midnight is modeled by $d(t) = 1.25 \cos \frac{\pi}{12}(t - 1) + 6$?

   F 4.75 m

   G 5.75 m

   H 7.25 m

   J 8.25 m

Name ______________________ Date __________ Class __________

LESSON **14-1**

Reading Strategy
Use a Visual Map

The form of the sine or cosine function determines how the graph varies. Use the graph of sin$x$ as a reference.

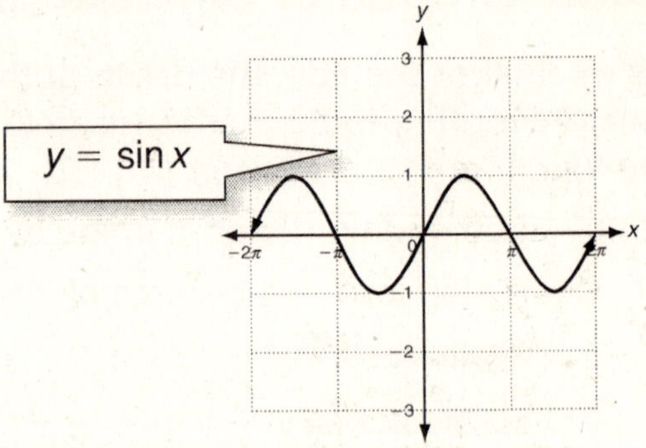

The absolute value of a, $|a|$, tells you the value of the amplitude of the function.
$y = \mathbf{2}\sin x$

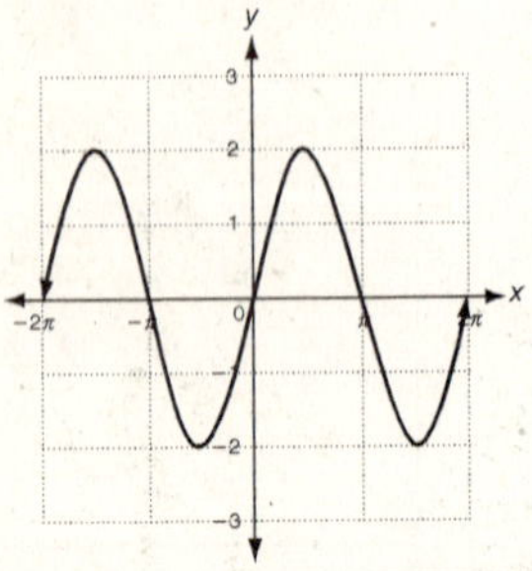

The value of the vertical shift, or vertical translation, is represented by k.
$y = \sin(x) + \mathbf{2}$

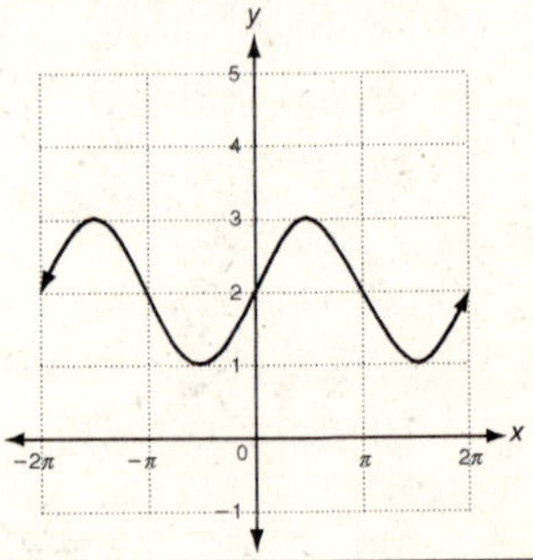

$$y = a\sin b(x - h) + k$$

From the variable b you can find the period, $\frac{2\pi}{|b|}$. The period tells you the length of one complete cycle.
$y = \sin(\mathbf{2}x)$

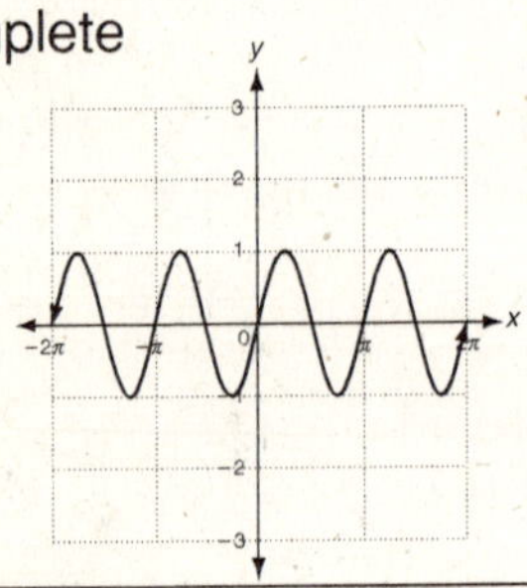

The value of the phase shift, or horizontal translation, is represented by h.
$y = \sin(x - \mathbf{2})$

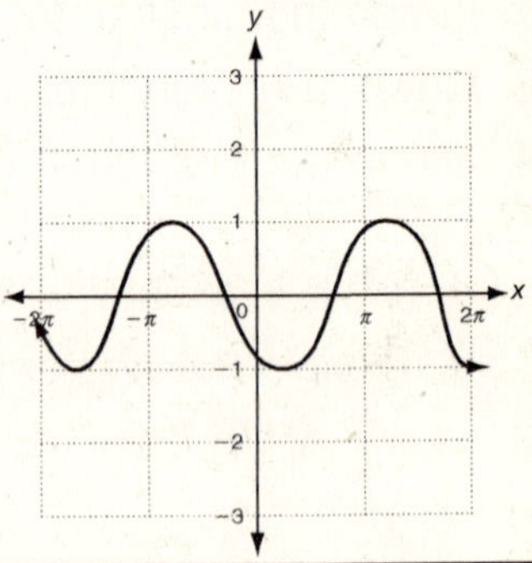

Answer each question.

1. Describe how the period of $f(x) = \sin x$ compares to the period of $g(x) = 4\sin 3x$.

____

2. Describe how the period of $f(x) = \cos x$ compares to the period of $g(x) = \frac{1}{2}\cos(x - 3)$.

____

3. Describe the phase shift for the function $g(x) = 2\sin 3(x - 8) + 5$.

____

4. What are the amplitude and period for the function $g(x) = 3\sin\frac{1}{2}(x + 2) - 1$?

____

5. What are the phase shift and the vertical shift for the function $g(x) = \cos 4(x + 5) + 3$?

____

Name ______________________ Date ____________ Class ____________

LESSON 14-2 Practice A

Graphs of Other Trigonometric Functions

Graph each function. Use $f(x) = \tan x$ as a guide. Identify the period, x-intercepts, and asymptotes.

1. $p(x) = \tan(3x)$

   a. Identify the period. ______________________

   b. Identify the x-intercepts. ______________________

   c. Identify the asymptotes. ______________________

   d. Use the information to graph the function.

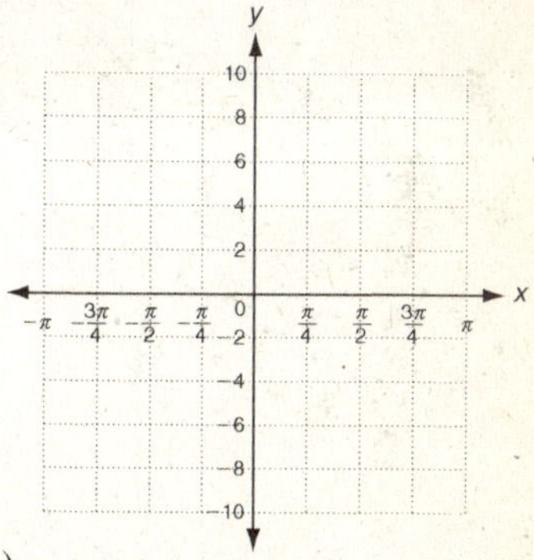

2. $q(x) = \tan\left(\frac{x}{2}\right)$

____

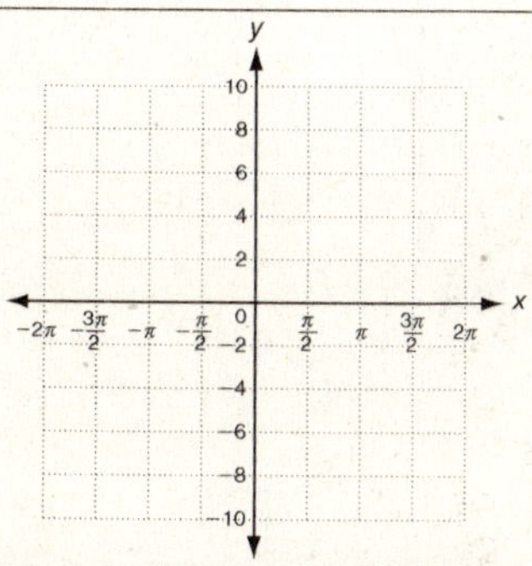

3. $k(x) = \tan(\pi x)$

____

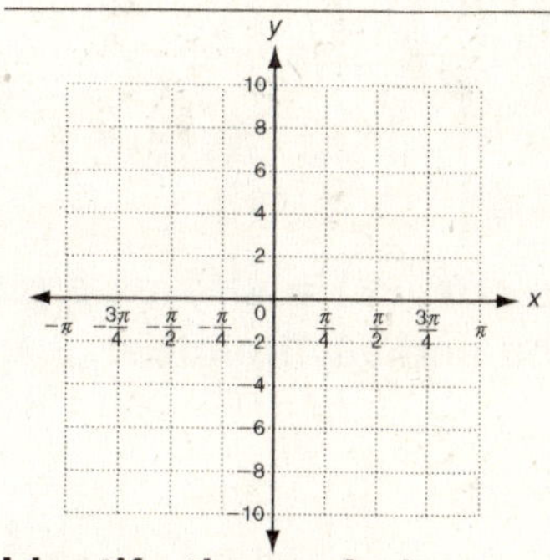

Graph each function. Use $f(x) = \cot x$ as a guide. Identify the period, x-intercepts, and asymptotes.

4. $h(x) = 2\cot\left(\frac{x}{2}\right)$

   a. Identify the period. ______________________

   b. Identify the x-intercepts. ______________________

   c. Identify the asymptotes. ______________________

   d. Use the information to graph the function.

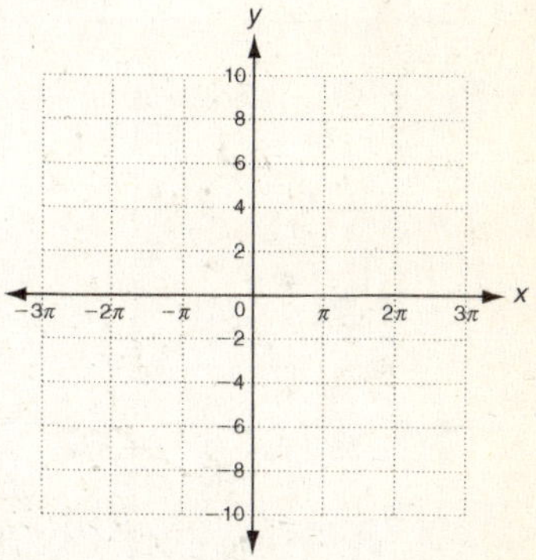

5. $b(x) = \frac{1}{2}\cot(2x)$

____

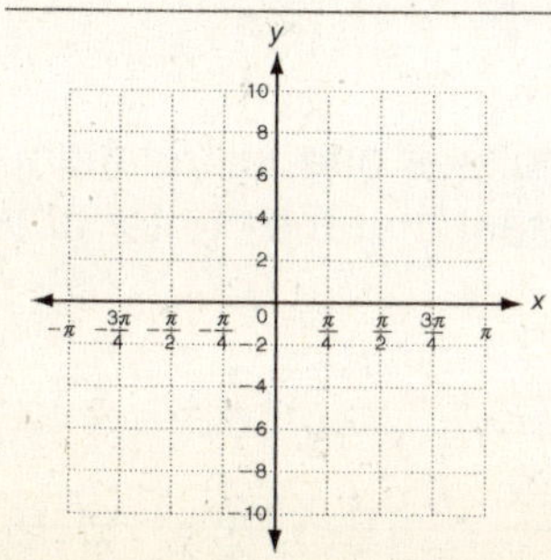

6. $b(x) = \frac{1}{2}\cot(x)$

____

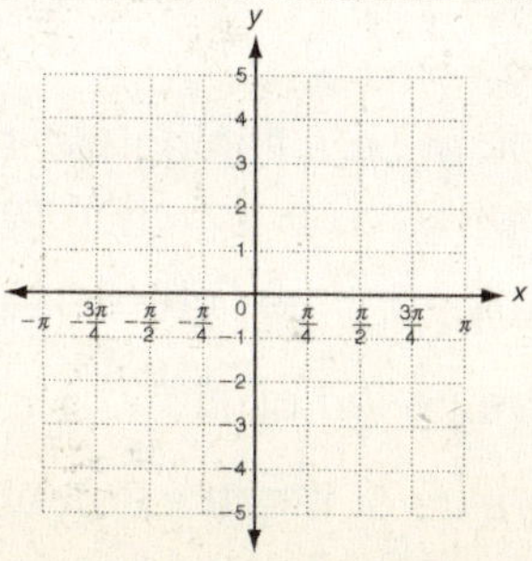

Name ______________________ Date ____________ Class ____________

LESSON 14-2

Practice B

Graphs of Other Trigonometric Functions

Using $f(x) = \tan x$ and $f(x) = \cot x$ as a guide, graph each function. Identify the period, x-intercepts, and asymptotes.

1. $g(x) = 2\tan\frac{\pi x}{2}$

____

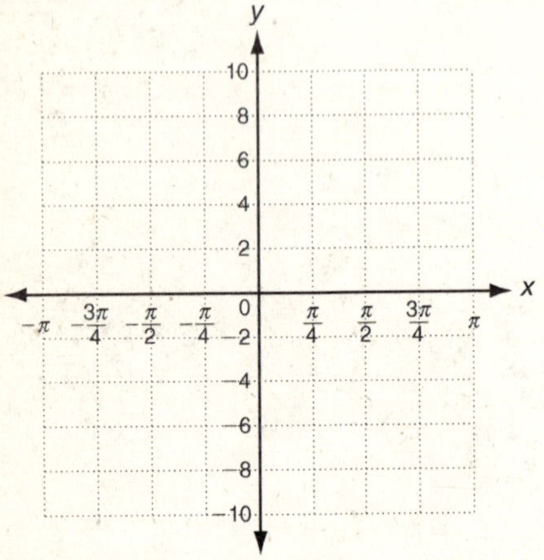

2. $t(x) = \frac{3}{4}\cot(x)$

____

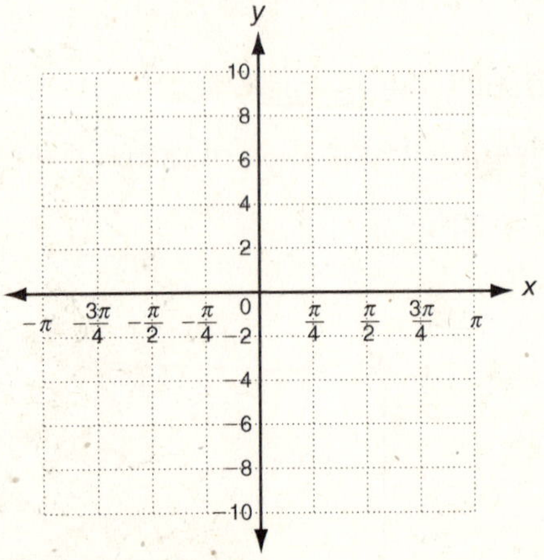

Using $f(x) = \cos x$ or $f(x) = \sin x$ as a guide, graph each function. Identify the period and asymptotes.

3. $k(x) = \sec\frac{x}{2}$

____

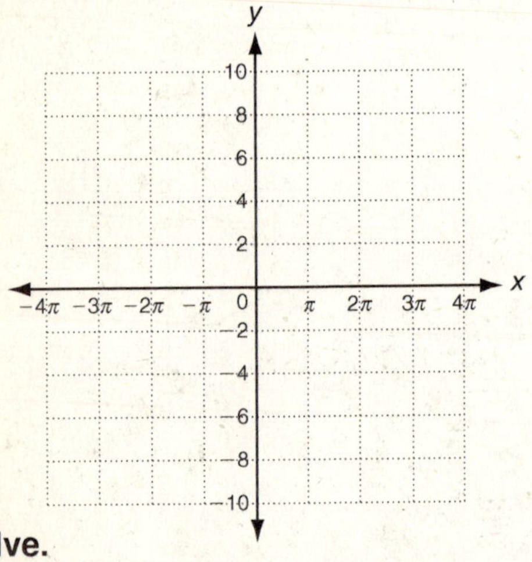

4. $q(x) = \frac{1}{2}\csc(2x)$

____

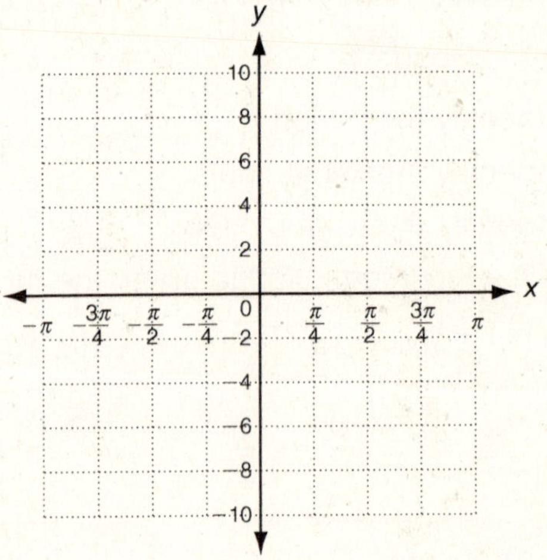

Solve.

5. The rotating light on a lighthouse is 400 feet from a cliff and completes one full rotation every 10 seconds. The equation representing the distance, a, in feet that the center of the circle of light is from the lighthouse is $a(t) = 400\sec\left(\frac{\pi t}{5}\right)$.

   a. What is the period of $a(t)$? ________________________

   b. Find the value of the function at $t = 10$. ________________________

Name ______________________ Date __________ Class __________

LESSON 14-2

Practice C

Graphs of Other Trigonometric Functions

Using $f(x) = \tan x$ or $f(x) = \cot x$ as a guide, graph each function. Identify the period, x-intercepts, and asymptotes.

1. $k(x) = \frac{1}{4}\tan(\pi x)$

____

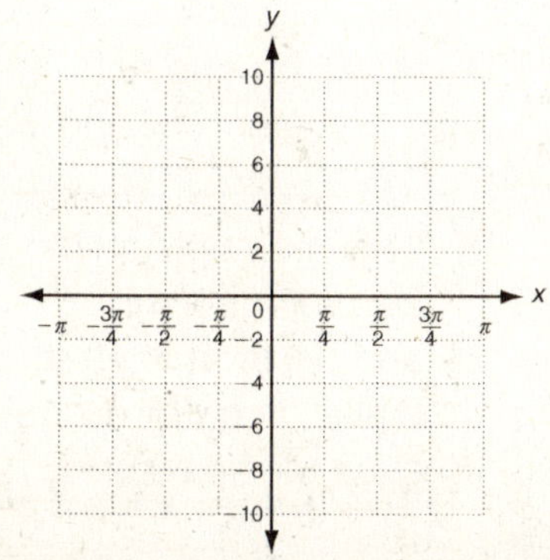

2. $c(x) = \cot\left(\frac{2x}{3}\right)$

____

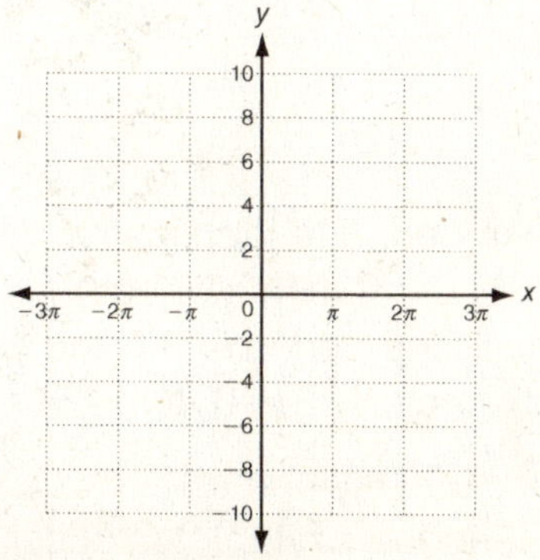

Using $f(x) = \cos x$ or $f(x) = \sin x$ as a guide, graph each function. Identify the period and asymptotes.

3. $k(x) = \sec\frac{x}{4}$

____

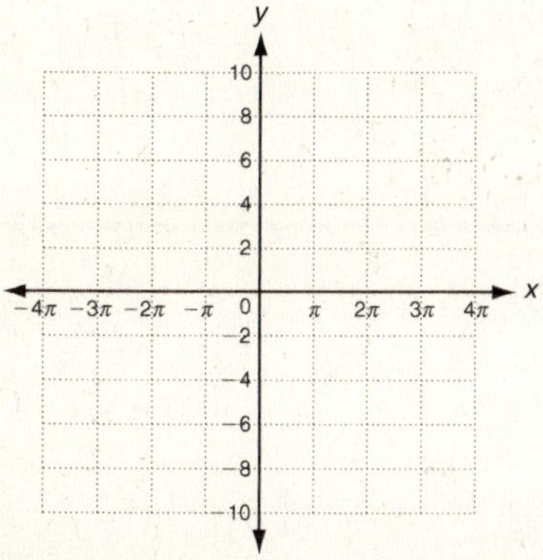

4. $q(x) = 2\csc(2x)$

____

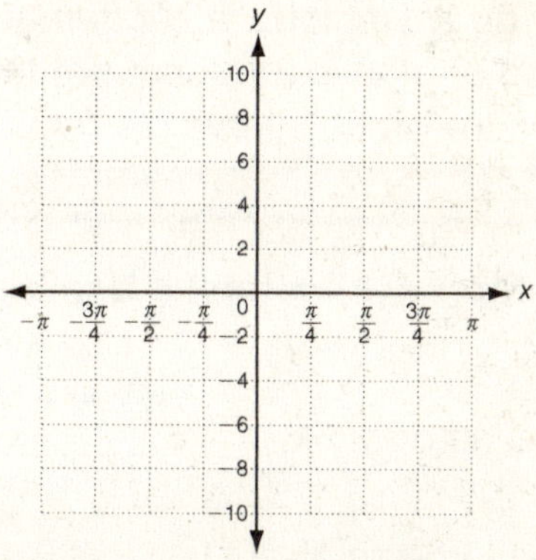

Solve.

5. A strobe light is located in the center of a square dance room. The rotating light is 40 feet from each of the 4 walls and completes one full rotation every 6 seconds. The equation representing the distance, d, in feet that the center of the circle of light is from the light source is $d(t) = \left|40\sec\left(\frac{\pi t}{3}\right)\right|$.

a. What is the period of $d(t)$? ______________________

b. Find the value of the function at $t = 2.5$. ______________________

Name ______________________ Date __________ Class __________

LESSON 14-2

Reteach

Graphs of Other Trigonometric Functions

Transformations of the tangent function change the period and/or asymptotes of the graph.

For $y = a\tan bx$: Note: x is measured in radians.

- the period is $\frac{\pi}{|b|}$,
- the asymptotes are located at $x = \frac{\pi}{2|b|} + \frac{\pi n}{|b|}$, where n is an integer.

Use the graph of $f(x) = \tan x$ to sketch the graph of $g(x) = \tan\frac{1}{2}x$.

Step 1 Find b to identify the period.

$b = \frac{1}{2}$, and $\frac{\pi}{|b|} = \frac{\pi}{\left|\frac{1}{2}\right|} = 2\pi$, so the period is 2π.

There will be one full cycle between $-\pi$ and π.

Step 2 Use the period to identify the x-intercepts.
The first x-intercepts of both $f(x)$ and $g(x)$ occur at 0.
Because the period is 2π, the intercepts occur at $2\pi n$.
For example, -2π, 0, and 2π are x-intercepts.

The intercepts shift 2π radians right and occur at integer multiples of 2π or the even multiples of π.

Step 3 Identify the asymptotes.
$b = \frac{1}{2}$, so the asymptotes occur at $x = \frac{\pi}{2\left|\frac{1}{2}\right|} + \frac{\pi n}{\left|\frac{1}{2}\right|} = \pi + 2\pi n$.

Step 4 Graph $f(x) = \tan x$ and $g(x) = \tan\frac{1}{2}x$ on the same plane.

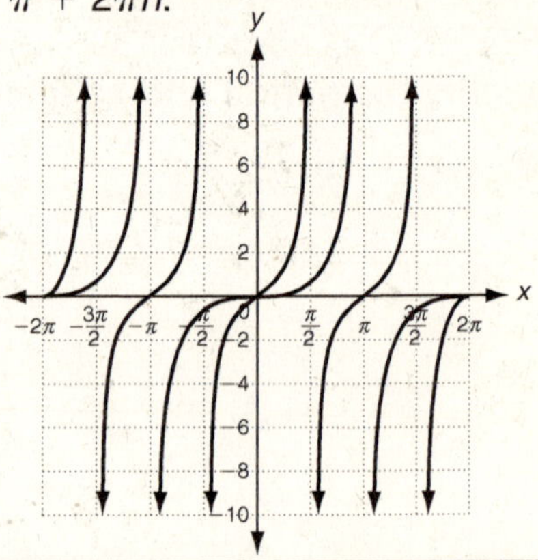

The x-intercepts of $g(x)$ are also x-intercepts of $f(x)$. The asymptotes are different. Because the period of $g(x)$ is 2π, there are fewer cycles of $g(x)$ in the same interval as $f(x)$.

Complete to graph $g(x) = \tan 2x$. Use the interval from $-\pi$ to π.

1. Find the period of $g(x)$. $\frac{\pi}{|b|} =$ ______________

2. Find the x-intercepts of $g(x)$.

____

3. Find the asymptotes of $g(x)$.

____

4. Sketch the graph of $f(x) = \tan x$. Then graph $g(x) = \tan 2x$.

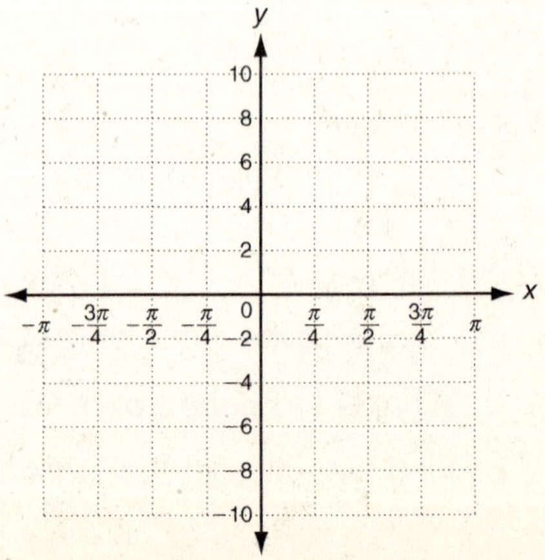

Name ______________________ Date __________ Class __________

LESSON 14-2

Reteach

Graphs of Other Trigonometric Functions (continued)

Transformations of the cotangent function are similar to transformations of the tangent function.

For $y = a\cot bx$:

- the period is $\frac{\pi}{|b|}$, — This is the same as for tangent transformations.
- the asymptotes are located at $x = \frac{\pi n}{|b|}$, where n is an integer.

Use the graph of $f(x) = \cot x$ to sketch the graph of $g(x) = \cot 2x$.

Step 1 Find b to identify the period.

$b = 2$, and $\frac{\pi}{|b|} = \frac{\pi}{|2|} = \frac{\pi}{2}$, so the period is $\frac{\pi}{2}$. — There will be one full cycle between 0 and $\frac{\pi}{2}$.

Step 2 Use the period to identify the x-intercepts.

The first x-intercept of $f(x)$ is $\frac{\pi}{2}$.

The first x-intercept of $g(x)$ occurs at $\frac{\pi}{2} \cdot \frac{1}{2}$, or $\frac{\pi}{4}$. — The x-intercepts shift $\frac{\pi}{2}$ radians right.

From $-\pi$ to π the x-intercepts are $-\frac{3\pi}{4}$, $-\frac{\pi}{4}$, $\frac{\pi}{4}$, and $\frac{3\pi}{4}$.

Step 3 Identify the asymptotes.

$b = 2$, so the asymptotes occur at $x = \frac{\pi n}{|2|} = \frac{\pi n}{2}$. — These are the multiples of $\frac{\pi}{2}$.

Step 4 Graph $g(x) = \cot 2x$.

- Sketch the asymptotes.
- Plot the intercepts.
- Then use $f(x) = \cot x$ as a guide to sketch the graph of $g(x) = \cot 2x$.

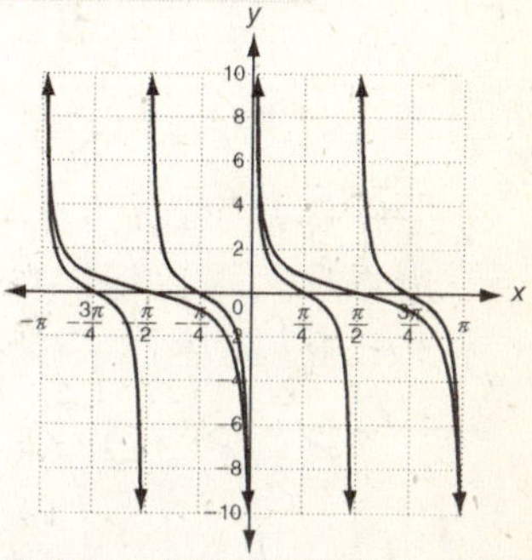

Complete to graph $g(x) = \cot\frac{1}{2}x$. Use the interval from -2π to 2π.

5. Find the period of $g(x)$. $\frac{\pi}{|b|} =$ ____________
6. Find the x-intercepts of $g(x)$. ____________
7. Where are the asymptotes of $g(x)$?

____

8. Sketch the graph of $g(x) = \cot\frac{1}{2}x$.

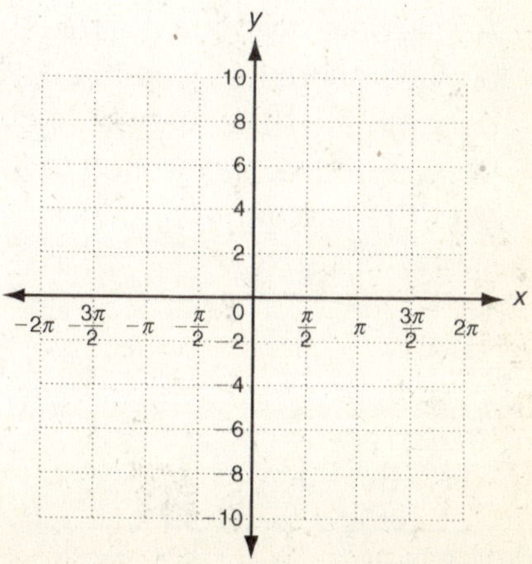

Name ______________________ Date ____________ Class ____________

LESSON 14-2

Challenge

Trigonometric Functions from the Graph

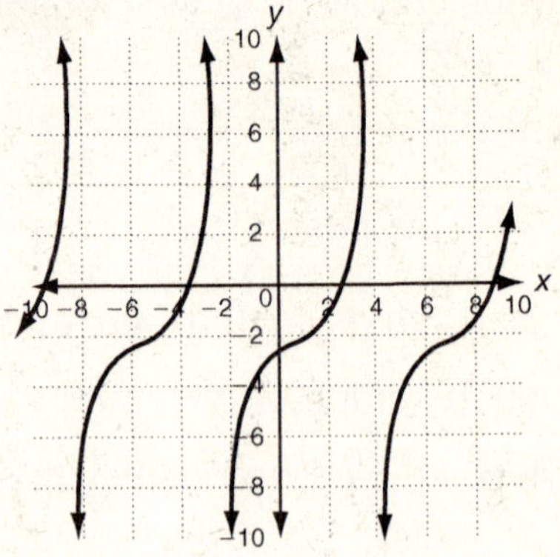

Consider the graph at right. It looks like a tangent graph with some transformations. The graph of $y = \tan x$ has the center point of one period pass through the origin. The center of the middle period of this graph passes through the point $(1, -2)$. This indicates that both a vertical translation and a phase shift are included in the equation of the function. The period is 2π and indicates a stretch in the period of the graph of the parent function, $y = \tan x$ by a factor of 2. Using these observations, an equation for the function is $y = \tan(0.5(x - 1)) - 2$. Due to the periodic nature of the tangent function, other equations are possible.

Write an equation for each graph below.

1.

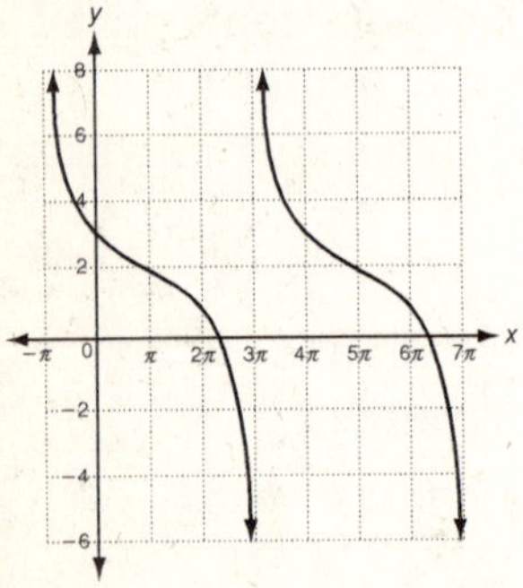

____

2.

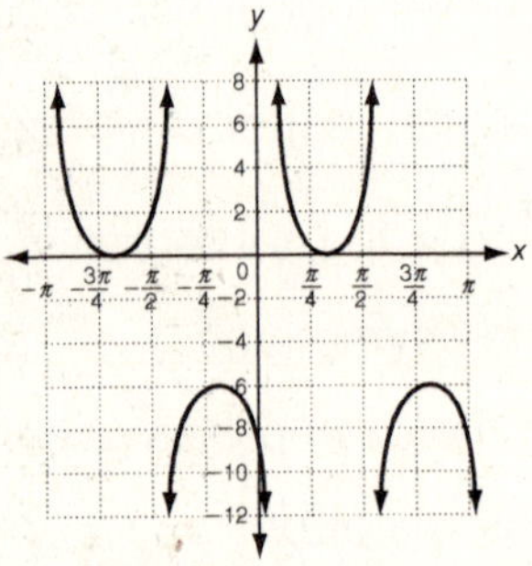

____

3.

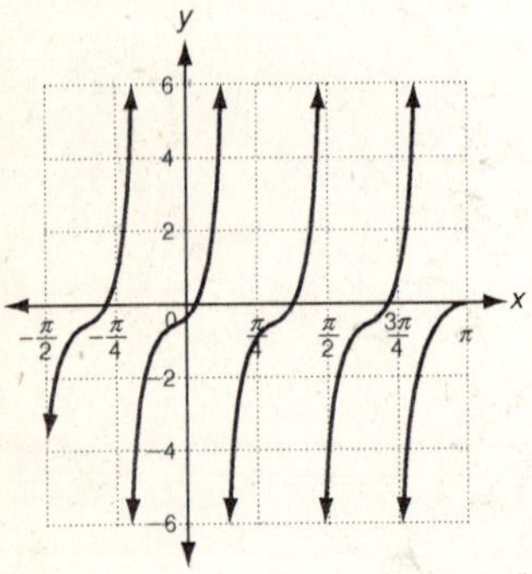

____

4.

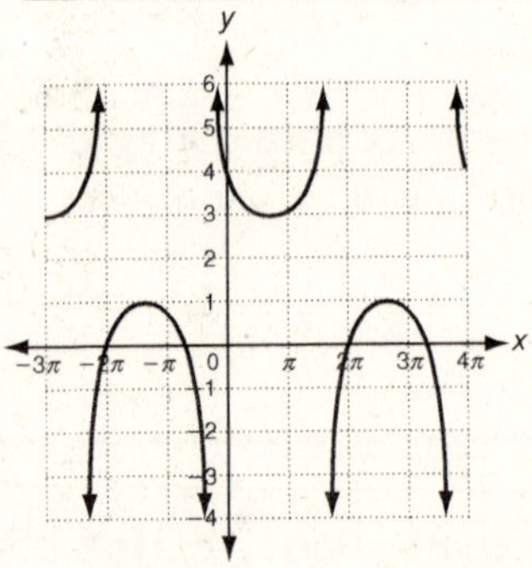

____

5. A helicopter flying in a straight line at a constant altitude of 2000 feet will pass directly over a playground. Let y be the distance from the playground to a point on the ground directly beneath the helicopter as shown in the diagram. Let x be the angle of elevation from a person on the playground to the helicopter. Write y as a function of x and sketch a graph over the interval from 0 to π.

____

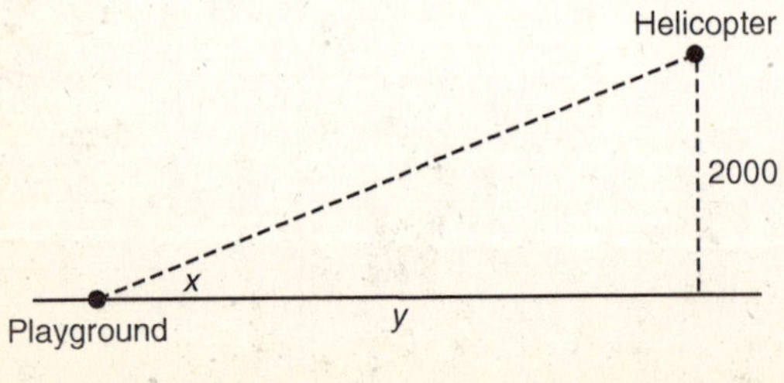

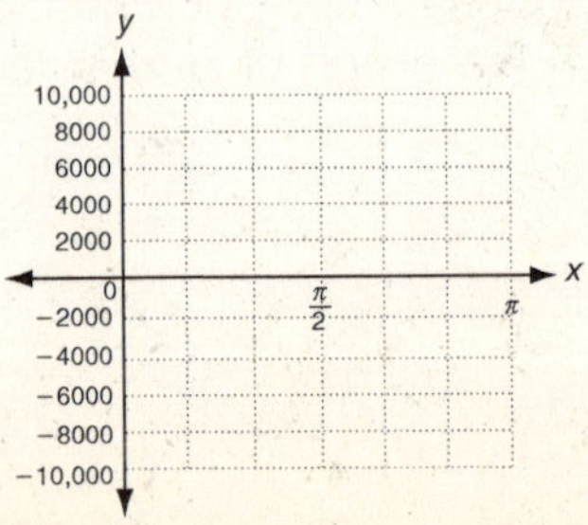

Name ______________________ Date ____________ Class ____________

LESSON **14-2**

Problem Solving

Graphs of Other Trigonometric Functions

As the sign outside the community youth center rotates, the light on it traces a path on the center's wall. The distance in feet of the light to the point on the wall where it shines varies with time, *t*, and can be modeled by the function $g(t) = 9 \csc\left(\frac{3\pi}{4}t\right)$, where *t* is measured in seconds.

1. Identify the period of the function.

   a. What is the coefficient of *t* in the equation of the function? ____________

   b. Write and evaluate an expression for the period of the function. ____________

   c. Explain what the period of the function means in terms of the time for one rotation of the light.

   ____

   d. How does the graph of the function vary based on the period?

   ____

2. a. This function has asymptotes at points where what other function is equal to 0?

   ____

   b. What are the asymptotes of the function?

   ____

3. What is the significance of 9 in the equation for the function?

   ____

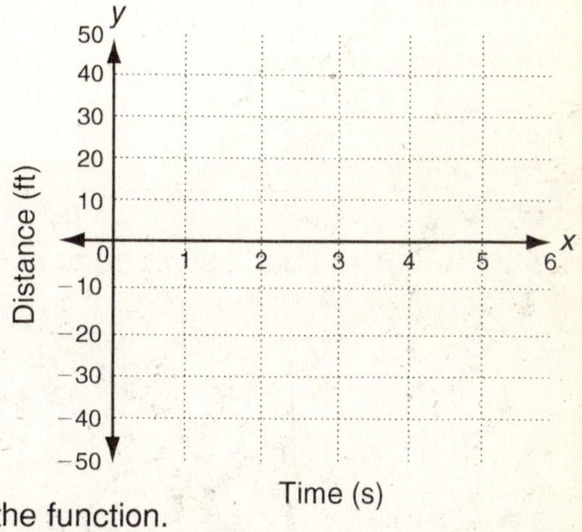

4. Graph the function using all the information about the function.

Toni changes the rotation rate of the light so that it rotates once every 4 seconds. Choose the letter for the best answer.

5. Which equation represents the distance of the light to the point on the wall where it shines, as a function of time, *t*?

   A $g(t) = 9 \csc\left(\frac{\pi}{4}t\right)$

   B $g(t) = 9 \csc\left(\frac{\pi}{2}t\right)$

   C $g(t) = 9 \csc(2\pi t)$

   D $g(t) = 9 \csc(4\pi t)$

6. Which pair of parameters describes this new function?

   F period: 4 s; vertical asymptotes at $\pm 2n$ s

   G period: 2 s; vertical asymptotes at $\pm n$ s

   H period: 2 s; horizontal asymptotes at $\pm 2n$ s

   J period: 4 s; horizontal asymptotes at $\pm n$ s

Name ________________________ Date ____________ Class ____________

LESSON **14-2**

Reading Strategy

Use Graphic Aids

Tangent and cotangent graphs have different characteristics.

In **tangent graphs** of the form $y = a\tan bx$, the period is $\frac{\pi}{|b|}$ and the asymptotes are at $x = \frac{\pi}{2|b|} + \frac{\pi n}{|b|}$.

$y = \tan x$

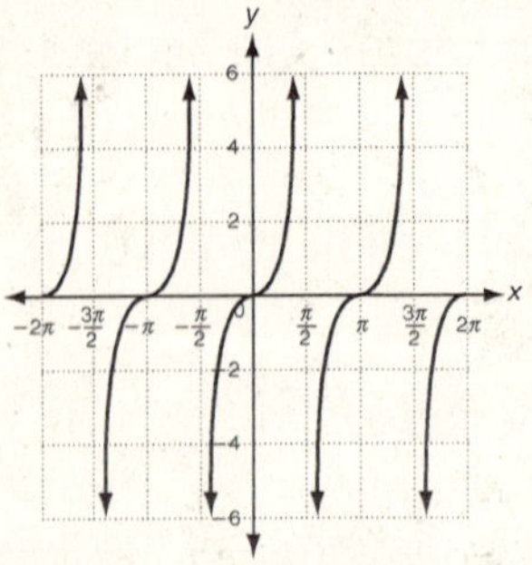

In **cotangent graphs** of the form $y = a\cot bx$, the period is $\frac{\pi}{|b|}$ and the asymptotes are at $x = \frac{\pi n}{|b|}$.

$y = \cot x$

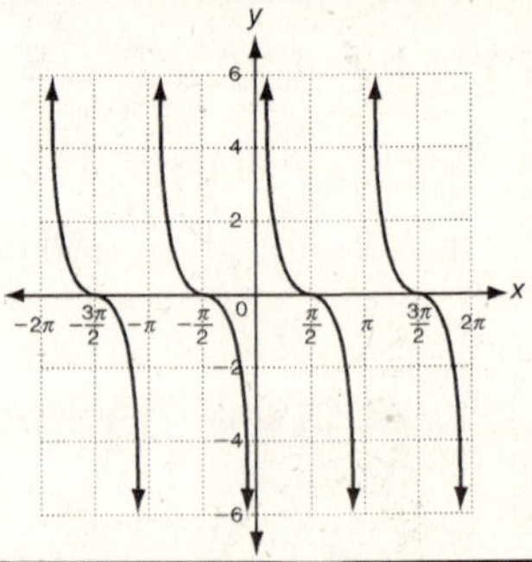

Determine whether each graph is a tangent or cotangent function.

1.

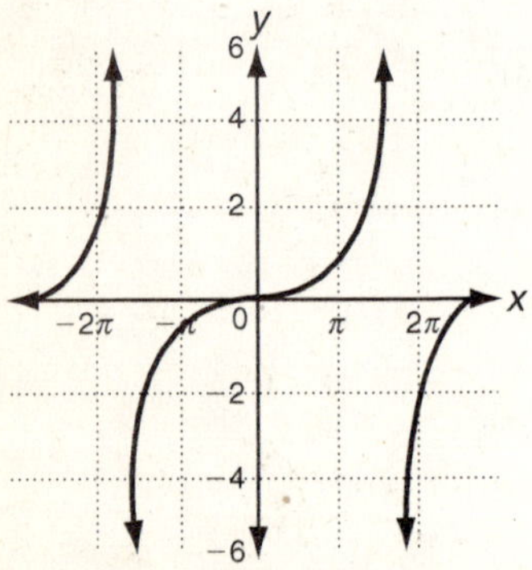

____

2.

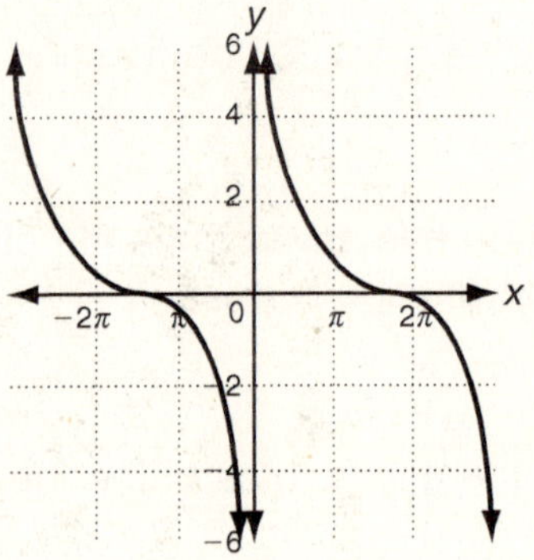

____

Identify the period and asymptotes of each graph or function.

3.

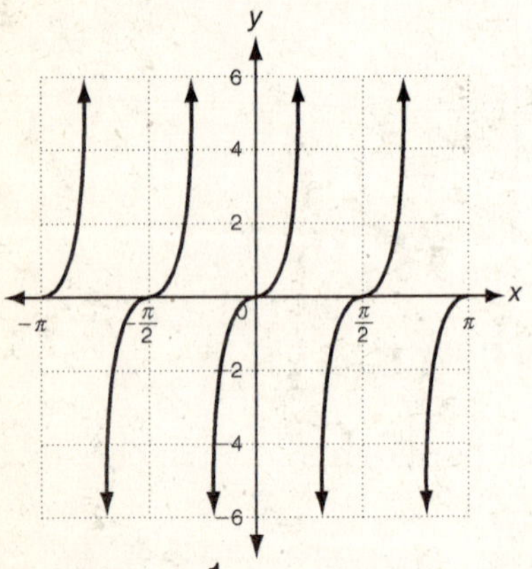

a. Identify the period.

____

b. Identify the asymptotes.

____

4. $g(x) = \tan\frac{1}{2}x$

a. Identify the period. ____________

b. Identify the asymptotes. ____________

5. $g(x) = \cot 2x$

a. Identify the period. ____________

b. Identify the asymptotes. ____________

Name ______________________ Date __________ Class __________

LESSON 14-3

Practice A

Fundamental Trigonometric Identities

Prove each trigonometric identity.

1. $\dfrac{\sin\theta}{1 - \cos^2\theta} = \csc\theta$

   a. Modify the left-hand side. Replace 1 with a known identify. ____________

   b. Simplify the denominator. ____________

   c. Keep simplifying and substituting identities until the left side matches the right side. ____________

2. $\cos^4\theta + \cos^2\theta\sin^2\theta + \sin^2\theta = 1$

3. $\cot\theta = \dfrac{\csc\theta}{\sec\theta}$

____________ ____________

Rewrite each expression in terms of cosθ. Then simplify.

4. $\dfrac{\cot\theta}{2\csc\theta}$

   a. Rewrite each function in terms of $\sin\theta$ or $\cos\theta$. ____________

   b. Rewrite the fraction in simplest form. ____________

5. $\sin\theta\cot\theta$

6. $\dfrac{\cos\theta - \cos^2\theta}{\sin^2\theta}$

7. $\dfrac{\sin^2\theta + 2\cos^2\theta}{\sec^2\theta - \tan^2\theta}$

____________ ____________ ____________

Solve.

8. Use the equation $mg\sin\theta = \mu mg\cos\theta$ to determine the angle at which a desk can be tilted before a paperback book on the desk begins to slide. Assume $\mu = 1.11$. ____________

Name ______________________ Date __________ Class __________

LESSON 14-3

Practice B

Fundamental Trigonometric Identities

Prove each trigonometric identity.

1. $\sin^2\theta + \sin^2\theta \cot^2\theta = 1$

2. $\cot^2\theta \cos^2\theta = \cot^2\theta - \cos^2\theta$

3. $\tan^2\theta - \tan^2\theta \sin^2\theta = \sin^2\theta$

4. $\dfrac{\sin\theta + \cos\theta}{\sin\theta \cos\theta} = \sec\theta + \csc\theta$

Rewrite each expression in terms of cos θ. Then simplify.

5. $2\sin\theta \cos\theta \cot\theta$

6. $\dfrac{1 + \cot\theta}{\cot\theta(\sin\theta + \cos\theta)}$

7. $\cos^4\theta - \sin^4\theta + \sin^2\theta$

Solve.

8. Use the equation $mg \sin\theta = \mu mg \cos\theta$ to determine the angle at which a waxed wood block on an inclined plane of wet snow begins to slide. Assume $\mu = 0.17$.

Name ______________________ Date __________ Class __________

LESSON 14-3

Practice C

Fundamental Trigonometric Identities

Prove each trigonometric identity.

1. $\dfrac{\cos\theta}{1+\sin\theta} = \dfrac{1-\sin\theta}{\cos\theta}$

2. $\tan\theta\sec\theta = \dfrac{\sec\theta+\tan\theta}{\cos\theta+\cot\theta}$

______________________ ______________________

3. $\tan^2\theta - \sin^2\theta = \tan^2\theta\sin^2\theta$

4. $1 + \cot^4\theta = \csc^4\theta - 2\cot^2\theta$

______________________ ______________________

Rewrite each expression in terms of $\sin\theta$ and $\cos\theta$. Then simplify.

5. $\dfrac{\cot\theta}{\sin\theta}$

6. $\cot^2\theta - \cos\theta\cot^2\theta$

7. $\tan^2\theta + 4\sec^2\theta + 1$

______________ ______________ ______________

Solve.

8. Alan is using the equation $mg\sin\theta = \mu mg\cos\theta$ to determine the coefficient of friction, μ, between a flat rock and a metal ramp. Find μ to the nearest hundredth if the rock begins to slide at 19°.

____

Name ______________________ Date ___________ Class ___________

LESSON 14-3

Reteach

Fundamental Trigonometric Identities

To prove a trigonometric identity, use the fundamental identities to make one side of the equation resemble the other side.

Reciprocal and Ratio Identities		
$\csc\theta = \dfrac{1}{\sin\theta}$	$\sec\theta = \dfrac{1}{\cos\theta}$	$\cot\theta = \dfrac{1}{\tan\theta}$
$\tan\theta = \dfrac{\sin\theta}{\cos\theta}$		$\cot\theta = \dfrac{\cos\theta}{\sin\theta}$

Negative-Angle Identities	Pythagorean Identities
$\sin(-\theta) = -\sin\theta$	$\sin^2\theta + \cos^2\theta = 1$
$\cos(-\theta) = \cos\theta$	$\tan^2\theta + 1 = \sec^2\theta$
$\tan(-\theta) = -\tan\theta$	$1 + \cot^2\theta = \csc^2\theta$

As long as you know this identity, you can divide to derive the other two.

Prove: $\tan\theta\sin\theta = \sec\theta - \cos\theta$

Try the right side since you can use reciprocal identities

$\tan\theta\sin\theta = \sec\theta - \cos\theta$	*Modify the right side.*
$= \dfrac{1}{\cos\theta} - \cos\theta$	*Definition of secant*
$= \dfrac{1 - \cos^2\theta}{\cos\theta}$	*Subtract. Use a common denominator.*
$= \dfrac{\sin^2\theta}{\cos\theta}$	*Pythagorean identity*
$= \dfrac{\sin\theta}{\cos\theta}\sin\theta$	*Rewrite the product.*
$= \tan\theta\sin\theta$	*Definition of tangent*

$\sin^2\theta + \cos^2\theta = 1$, so $\sin^2\theta = 1 - \cos^2\theta$.

Write the missing step or reason to prove each trigonometric identity.

1. $\cot(-\theta) = -\cot\theta$ Modify the left side.

$\dfrac{\cos(-\theta)}{\sin(-\theta)} = -\cot\theta$ ______________________

$\dfrac{\cos\theta}{\sin(-\theta)} = -\cot\theta$ Negative-angle identity

$\dfrac{\cos\theta}{-\sin\theta} = -\cot\theta$ ______________________

$-\cot\theta = -\cot\theta$ ______________________

2. $\sec\theta\cot\theta = \csc\theta$ Modify the left side.

$\left(\dfrac{1}{\cos\theta}\right)\left(\dfrac{\cos\theta}{\sin\theta}\right) = \csc\theta$ ______________________

______________ Simplify.

$\csc\theta = \csc\theta$ ______________________

Name ______________________ Date ___________ Class ___________

LESSON 14-3

Reteach

Fundamental Trigonometric Identities (continued)

Rewriting a trigonometric expression is similar to proving a trigonometric identity. Use the fundamental identities to rewrite the expression.

Trigonometric Identities		
$\csc\theta = \dfrac{1}{\sin\theta}$	$\sec\theta = \dfrac{1}{\cos\theta}$	$\cot\theta = \dfrac{1}{\tan\theta}$
$\tan\theta = \dfrac{\sin\theta}{\cos\theta}$	$\cot\theta = \dfrac{\cos\theta}{\sin\theta}$	
$\sin(-\theta) = -\sin\theta$	$\sin^2\theta + \cos^2\theta = 1$	
$\cos(-\theta) = \cos\theta$	$\tan^2\theta + 1 = \sec^2\theta$	
$\tan(-\theta) = -\tan\theta$	$1 + \cot^2\theta = \csc^2\theta$	

Rewrite $\dfrac{\cos^2\theta}{1 - \sin\theta}$ in terms of $\sin\theta$.

$\dfrac{\cos^2\theta}{1 - \sin\theta}$

$\sin^2\theta + \cos^2\theta = 1$, so $\cos^2\theta = 1 - \sin^2\theta$.

$\dfrac{1 - \sin^2\theta}{1 - \sin\theta}$ *Pythagorean identity*

$\dfrac{(1 - \sin\theta)(1 + \sin\theta)}{1 - \sin\theta}$ *Factor the numerator.*

Think: Let $a = \sin\theta$. $1 - a^2 = (1 - a)(1 + a)$

$\dfrac{\cancel{(1 - \sin\theta)}(1 + \sin\theta)}{\cancel{1 - \sin\theta}}$ *Simplify.*

$1 + \sin\theta$ The expression is in terms of $\sin\theta$.

Write the missing step or reason to rewrite each expression in terms of $\cos\theta$.

3. $(1 + \sec^2\theta)(1 - \sin^2\theta)$

$\left(1 + \dfrac{1}{\cos^2\theta}\right)(1 - \sin^2\theta)$ ______________________

$\left(1 + \dfrac{1}{\cos^2\theta}\right)(\cos^2\theta)$ ______________________

______________________ Multiply.

4. $\tan^2\theta$

______________________ Definition of tangent

$\dfrac{1 - \cos^2\theta}{\cos^2\theta}$ ______________________

______________________ Use the Distributive Property.

Name ______________________ Date ____________ Class ____________

LESSON 14-3

Challenge

Expressing One Function in Terms of Another

Equations that are true for all values of the variables for which they are defined are called *identities*. There are three groups of basic trigonometric identities.

Reciprocal Identities	Ratio Identities	Pythagorean Identities
$\csc\theta = \frac{1}{\sin\theta}$ with $\sin\theta \neq 0$	$\tan\theta = \frac{\sin\theta}{\cos\theta}$ with $\cos\theta \neq 0$	$\sin^2\theta + \cos^2\theta = 1$
$\sec\theta = \frac{1}{\cos\theta}$ with $\cos\theta \neq 0$	$\cot\theta = \frac{\cos\theta}{\sin\theta}$ with $\sin\theta \neq 0$	$\tan^2\theta + 1 = \sec^2\theta$
$\cot\theta = \frac{1}{\tan\theta}$ with $\tan\theta \neq 0$		$\cot^2\theta + 1 = \csc^2\theta$

You can use these identities to express one trigonometric function in terms of another.

To express $\tan\theta$ in terms of $\sin\theta$, begin with the ratio identity for $\tan\theta$.

$$\tan\theta = \frac{\sin\theta}{\cos\theta}$$

Now find $\cos\theta$ in terms of $\sin\theta$. Use the appropriate Pythagorean identity, and solve for $\cos\theta$.

$$\sin^2\theta + \cos^2\theta = 1$$
$$\cos^2\theta = 1 - \sin^2\theta$$
$$\cos\theta = \pm\sqrt{1 - \sin^2\theta}$$

Return to the ratio identity and substitute for $\cos\theta$.

$$\tan\theta = \frac{\sin\theta}{\pm\sqrt{1 - \sin^2\theta}}$$

Express the given function in terms of $\cos\theta$.

1. $\tan\theta$ ____________

2. $\cot\theta$ ____________

3. $\csc\theta$ ____________

Express the given function in terms of $\tan\theta$.

4. $\cot\theta$ ____________

5. $\sin\theta$ ____________

6. $\cos\theta$ ____________

Express the given function in terms of $\csc\theta$.

7. $\cos\theta$ ____________

8. $\tan\theta$ ____________

9. $\sec\theta$ ____________

Express the given function in terms of $\sec\theta$.

10. $\tan\theta$ ____________

11. $\sin\theta$ ____________

12. $\cot\theta$ ____________

Name ______________________ Date ____________ Class ____________

LESSON 14-3

Problem Solving

Fundamental Trigonometric Identities

The advertisement for a new shoe promises runners less slip. The coefficient of friction (μ) between concrete and a new material used in the sole of this shoe is 1.5. The force of friction that causes slip is equal to $mg \sin \theta$, where m is a runner's mass and g is the acceleration due to gravity. The force that prevents slip is $\mu mg \cos\theta$.

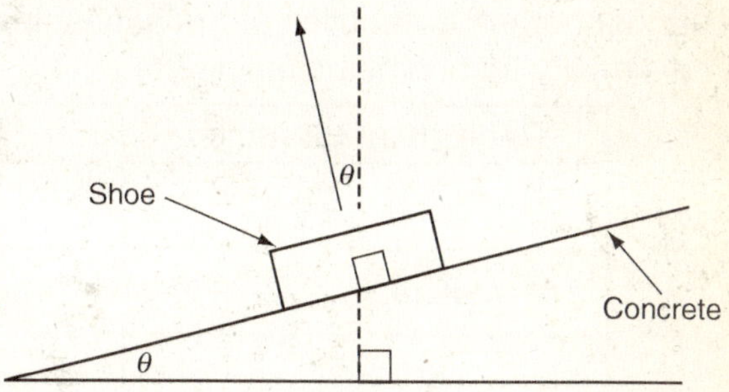

1. Mukisa wants to know what this new shoe will do for his performance.

   a. At the instant of slip, the force that causes slip is equal to the force that prevents it. Write an equation to show this relationship. ____________

   b. Rewrite the equation with trigonometric expressions on one side of the equation, and substitute the coefficient of friction for the new material. ____________

   c. Use a trigonometric identity to rewrite the equation using only the tangent function. ____________

   d. Solve for θ, the angle at which the new shoe will start to slip. ____________

2. Mukisa wonders how his old shoes compare to these new shoes. The label on the box of his old shoes states that the coefficient of friction is 1.3.

   a. Write an equation to find the angle at which a shoe with a coefficient of friction of 1.3 will slip. ____________

   b. At what angle will this shoe start to slip? ____________

   c. Suggest how the new shoes might improve Mukisa's performance.

____

An advertisement states, "Use this wax and your skis will slide better than silk on silk!" The coefficient of friction for silk on silk is 0.25. The coefficient of friction for waxed wood on wet snow is 0.14. The equation $mg \sin\theta = \mu mg \cos\theta$ can be used to find the angle at which a material begins to slide. Choose the letter for the best answer.

3. Which expression gives the angle at which silk begins to slide on silk?

   A $\cos^{-1}(0.25)$

   B $\sin\left(\frac{1}{0.25}\right)$

   C $\tan\left(\frac{1}{0.25}\right)$

   D $\tan^{-1}(0.25)$

4. At what angle would a waxed wooden ski begin to slide on wet snow?

   F 2°

   G 8°

   H 12°

   J 14°

Name ______________________ Date __________ Class __________

LESSON 14-3

Reading Strategy
Identify Relationships

The trigonometric identities are relationships that we can use to rewrite and simplify trigonometric functions.

Reciprocal Identities	Ratio Identities
$\csc\theta = \dfrac{1}{\sin\theta}$	$\tan\theta = \dfrac{\sin\theta}{\cos\theta}$
$\sec\theta = \dfrac{1}{\cos\theta}$	$\cot\theta = \dfrac{\cos\theta}{\sin\theta}$
$\cot\theta = \dfrac{1}{\tan\theta}$	

Answer each question.

1. Which identities are like inverse relationships? Explain.

____

2. Rewrite each as its reciprocal identity.

   a. sec85° ______________

   b. csc150° ______________

   c. cot22° ______________

3. Rewrite each as its ratio identity.

   a. tan340° ______________

   b. $\dfrac{1}{\tan 230°}$ ______________

   c. $\dfrac{1}{\cot 45°}$ ______________

Rewrite each expression (a) using reciprocal identities, and then (b) simplify using ratio identities.

4. $\csc\theta - \cot\theta$ a. ______________ b. ______________

5. $\sec\theta + \dfrac{1}{\cot\theta}$ a. ______________ b. ______________

6. $\cot\theta + 2(\csc\theta)$ a. ______________ b. ______________

Rewrite each expression in terms of sine and cosine using reciprocal identities and ratio identities.

7. $\dfrac{1}{\cot\theta}$ ______________________

8. $\dfrac{1}{\csc\theta} + \tan\theta$ ______________________

9. $\dfrac{1}{\tan\theta} + \csc\theta$ ______________________

10. $\dfrac{1}{\sec\theta} + \tan\theta$ ______________________

Name ______________________ Date __________ Class __________

LESSON 14-4

Practice A

Sum and Difference Identities

Find the exact value of each expression.

1. $\cos 210°$

   a. Write 210° as a sum or difference of two known trigonometric values. ______________

   b. Substitute the values into the correct sum or difference identity.

   ____

   c. Evaluate and simplify. ______________

2. $\cos 315°$ ______________

3. $\sin 105°$ ______________

4. $\tan 15°$ ______________

5. $\tan \frac{5\pi}{12}$ ______________

6. $\sin \frac{5\pi}{3}$ ______________

7. $\cos \frac{2\pi}{3}$ ______________

Let $\sin A = \frac{12}{13}$ with $90° \le A \le 180°$ and $\tan B = -\frac{4}{3}$ with $270° \le B \le 360°$. Find each value.

8. $\tan(A + B)$

   a. Use $\sin A = \frac{12}{13}$ to find $\tan A$. ______________

   b. Substitute known values into the angle sum identity $\tan(A + B) = \frac{\tan A + \tan B}{1 - \tan A \tan B}$.

   ____

   c. Simplify. ______________

9. $\tan(A - B)$ ______________

10. $\sin(A + B)$ ______________

11. $\cos(A - B)$ ______________

Solve.

12. The displacement, y, of a mass attached to a spring is modeled by $y(t) = 9.6\cos\left(\frac{2\pi}{6}(t) - \frac{\pi}{2}\right)$, where t is the time in seconds.

   a. What are the amplitude and period of the function? ______________

   b. Write the displacement using only the sine function. ______________

   c. What is the displacement of mass when $t = 6$? ______________

Name ______________________ Date __________ Class __________

LESSON 14-4

Practice B

Sum and Difference Identities

Find the exact value of each expression.

1. $\cos 120°$ ____________

2. $\sin 315°$ ____________

3. $\tan 255°$ ____________

4. $\tan\frac{7\pi}{6}$ ____________

5. $\sin\frac{\pi}{12}$ ____________

6. $\cos\frac{3\pi}{4}$ ____________

Prove each identity.

7. $\sin\left(x - \frac{3\pi}{2}\right) = \cos x$ ____________

8. $\cos\left(x - \frac{\pi}{2}\right) = \sin x$ ____________

Find each value if $\cos A = \frac{12}{13}$ with $0° \le A \le 90°$ and if $\sin B = \frac{8}{17}$ with $90° \le B \le 180°$.

9. $\sin(A + B)$ ____________

10. $\cos(A + B)$ ____________

11. $\tan(A + B)$ ____________

12. $\sin(A - B)$ ____________

13. $\cos(A - B)$ ____________

14. $\tan(A - B)$ ____________

Solve.

15. Find the coordinates, to the nearest hundredth, of the vertices of triangle *ABC* with *A*(1, 0), *B*(10, 0), and *C*(2, 6) after a 60° rotation about the origin.

a. Write the matrices for the rotation and for the points. ____________

b. Find the matrix product. ____________

c. Write the coordinates. ____________

16. A hill rises from the horizontal at a 15° angle. The road leading straight up the hill is 800 meters long. How much higher is the top of the hill than the base of the hill? ____________

Name ______________________ Date __________ Class __________

LESSON 14-4

Practice C

Sum and Difference Identities

Find the exact value of each expression.

1. $\cos 300°$

2. $\sin(-15°)$

3. $\tan 285°$

4. $\tan\frac{11\pi}{12}$

5. $\sin\frac{13\pi}{12}$

6. $\cos\left(-\frac{13\pi}{12}\right)$

Prove each identity.

7. $\tan(\pi - x) = -\tan x$

8. $\sin\left(\frac{\pi}{2} - x\right) = \cos x$

Find each value if $\sin A = \frac{12}{13}$ with $90° \le A \le 180°$ and if $\sin B = -\frac{7}{25}$ with $-90° \le B \le 0°$.

9. $\sin(A + B)$

10. $\cos(A + B)$

11. $\tan(A + B)$

12. $\sin(A - B)$

13. $\cos(A - B)$

14. $\tan(A - B)$

Solve.

15. Find the coordinates, to the nearest hundredth, of the vertices of triangle *PQR* with *P*(0, 1), *Q*(0, −4), and *R*(2, 5) after a 45° rotation about the origin.

16. Find the coordinates, to the nearest hundredth, of the vertices of triangle *FGH* with *F*(0, 10), *G*(10, 0), and *H*(1, −1) after a 75° rotation about the origin.

17. Two pilots take off from the same airport. Mason heads due south. Nancy heads 23° west of south. After 400 land miles, how far is Nancy from Mason's route?

Name ______________________ Date __________ Class __________

LESSON 14-4

Reteach

Sum and Difference Identities

You can use angle addition and subtraction identities to find the exact value of some trigonometric expressions.

Look for ways to use 30° or $\frac{\pi}{6}$, 45° or $\frac{\pi}{4}$, and 60° or $\frac{\pi}{3}$ to make the sum because exact values are known for these angles.

Sum Identities	Difference Identities
$\sin(A + B) = \sin A\cos B + \cos A\sin B$	$\sin(A - B) = \sin A\cos B - \cos A\sin B$
$\cos(A + B) = \cos A\cos B - \sin A\sin B$	$\cos(A - B) = \cos A\cos B + \sin A\sin B$
$\tan(A + B) = \frac{\tan A + \tan B}{1 - \tan A\tan B}$	$\tan(A - B) = \frac{\tan A - \tan B}{1 + \tan A\tan B}$

Find the exact value of cos105°.

$\cos 105° = \cos(60° + 45°)$ — *Think: 60° + 45° = 105°.* (Substitute: $A = 60°$ and $B = 45°$.)

$= \cos 60°\cos 45° - \sin 60°\sin 45°$ — *Use cos(A + B) identity.*

$= \frac{1}{2} \cdot \frac{\sqrt{2}}{2} - \frac{\sqrt{3}}{2} \cdot \frac{\sqrt{2}}{2}$ — *Evaluate.*

$= \frac{\sqrt{2}}{4} - \frac{\sqrt{6}}{4} = \frac{\sqrt{2} - \sqrt{6}}{4}$ — *Simplify.* (The value is negative. This makes sense since 105° lies in Quadrant II where cosine is negative.)

Find the exact value of $\sin\left(\frac{\pi}{12}\right)$.

$\sin\left(\frac{\pi}{12}\right) = \sin\left(\frac{\pi}{3} - \frac{\pi}{4}\right)$ — *Think: $\frac{\pi}{3} - \frac{\pi}{4} = \frac{\pi}{12}$.*

$= \sin\frac{\pi}{3}\cos\frac{\pi}{4} - \cos\frac{\pi}{3}\sin\frac{\pi}{4}$ — *Use sin(A − B) identity.* (Substitute: $A = \frac{\pi}{3}$ and $B = \frac{\pi}{4}$.)

$= \frac{\sqrt{3}}{2} \cdot \frac{\sqrt{2}}{2} - \frac{1}{2} \cdot \frac{\sqrt{2}}{2}$ — *Evaluate.*

$= \frac{\sqrt{6}}{4} - \frac{\sqrt{2}}{4} = \frac{\sqrt{6} - \sqrt{2}}{4}$ — *Simplify.*

Use the sum or difference identity to find the exact value of each expression.

1. $\cos(-15°) = \cos(30° - 45°)$

____

____

____

2. $\sin\left(\frac{7\pi}{12}\right) = \sin\left(\frac{\pi}{3} + \frac{\pi}{4}\right)$

____

____

____

Name ______________________ Date ___________ Class ___________

LESSON 14-4

Reteach

Sum and Difference Identities (continued)

You can use angle addition and subtraction identities to prove identities.
Use an identity to make one side of the equation resemble the other side.

Sum Identities	Difference Identities
$\sin(A + B) = \sin A \cos B + \cos A \sin B$	$\sin(A - B) = \sin A \cos B - \cos A \sin B$
$\cos(A + B) = \cos A \cos B - \sin A \sin B$	$\cos(A - B) = \cos A \cos B + \sin A \sin B$
$\tan(A + B) = \dfrac{\tan A + \tan B}{1 - \tan A \tan B}$	$\tan(A - B) = \dfrac{\tan A - \tan B}{1 + \tan A \tan B}$

Prove: $\tan(\pi + x) = \tan x$

$\tan(\pi + x) = \tan x$ — *Modify the left side.*

$\dfrac{\tan\pi + \tan x}{1 - \tan\pi \tan x} = \tan x$ — *Use tan(A + B) identity.*

Substitute: $A = \pi$ and $B = x$.

$\dfrac{0 + \tan x}{1 - (0)\tan x} = \tan x$ — *Evaluate. Think: $\tan\pi = 0$.*

$\dfrac{\tan x}{1} = \tan x$ — *Simplify.*

Prove: $\sin(\pi - x) = \sin x$

$\sin(\pi - x) = \sin x$ — *Modify the left side.*

$\sin\pi \cos x - \cos\pi \sin x = \sin x$ — *Use sin(A – B) identity.*

$(0)\cos x - (-1)\sin x = \sin x$ — *Evaluate. Think: $\sin\pi = 0$ and $\cos\pi = -1$.*

$0 + \sin x = \sin x$ — *Simplify.*

Write the missing steps or reasons to prove each identity.

3. $\sin\left(\frac{\pi}{2} + x\right) = \cos x$ — Modify the left side.

$\sin\left(\frac{\pi}{2}\right)\cos x + \cos\left(\frac{\pi}{2}\right)\sin x = \cos x$ — ______________________

______________________ — Evaluate.

______________________ — Simplify.

4. $\cos(\pi + x) = -\cos x$ — Modify the left side.

______________________ — Use $\cos(A + B)$ identity.

______________________ — ______________________

______________________ — ______________________

Name ______________________ Date ____________ Class ____________

LESSON **14-4**

Challenge

Proof of Identity

The diagram at right provides a geometric proof of the trigonometric identity $\sin(x + y) = \sin x \cos y + \cos x \sin y$. Angles x and y are located at the origin with angle y in standard position and the initial side of angle x located on the terminal side of angle y. The circle has $r = 1$. All other sides of the triangle are marked.

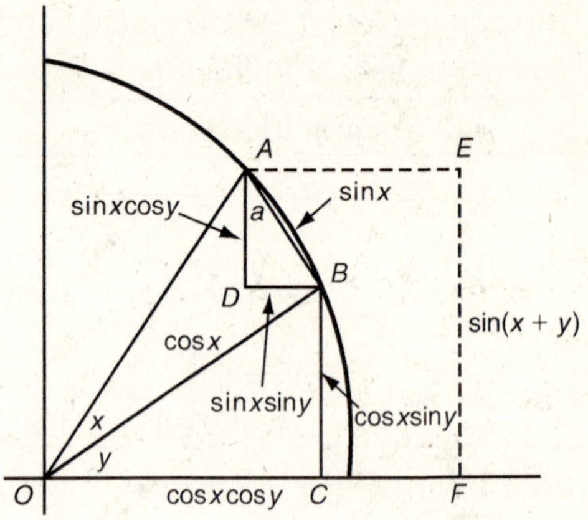

1. Using the properties of right triangles and the definitions of the trigonometric functions, justify the lengths of the sides of triangles AOB and BOC.

2. Prove that angle a in triangle ABD is equal in measure to angle y.

3. Using the properties of right triangles and the definitions of the trigonometric functions, justify the lengths of the sides of triangle ABD.

4. Now prove the identity for the sine of a sum, $\sin(x + y) = \sin x \cos y + \cos x \sin y$.

5. Use this information to prove the sum identity for cosines, $\cos(x + y) = \cos x \cos y - \sin x \sin y$.

Name ______________________ Date __________ Class __________

LESSON 14-4

Problem Solving

Sum and Difference Identities

Caitlin is designing a rotating graphic for her website. She explores what it will look like if it flashes in different positions.

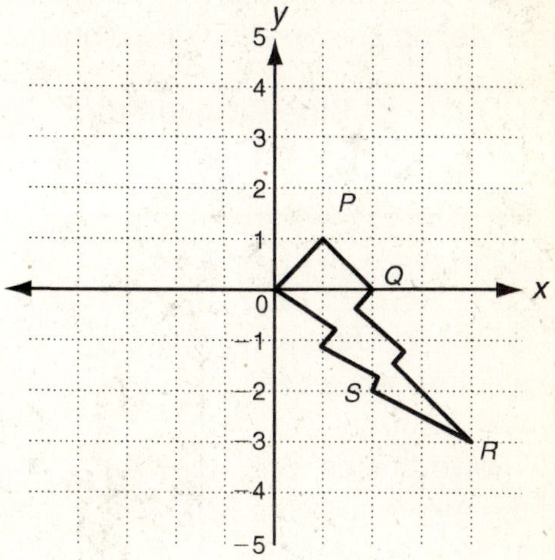

1. Use rotation matrices to find the coordinates of the figure *PQRS* after a 60° rotation about the origin.

   a. Write matrices for a 60° rotation and for the points *P*, *Q*, *R*, and *S* in the figure.

   $R_{60°} = \begin{bmatrix} & \end{bmatrix}$

   $PQRS = \begin{bmatrix} & \end{bmatrix}$

   b. Find the matrix product. $R_{60°} \times PQRS \approx \begin{bmatrix} & \end{bmatrix}$

   c. Name the coordinates of P', Q', R', S' to the nearest hundredth after a 60° rotation.

   ____

2. a. Write the matrix product that gives the coordinates of the figure after a rotation of 150°.

   $R_{150°} \times PQRS = \begin{bmatrix} & \end{bmatrix}$

   b. Name the coordinates of $P''Q''R''S''$ to the nearest hundredth.

   ____

3. Graph the two rotations on the plane above to see where Caitlin's rotating graphic will flash. Label R' and R''.

Choose the letter for the best answer.

4. Which expression gives the exact value of sin150°?

   A sin90°cos60° + cos90°sin60°

   B sin90°cos60° − cos90°sin60°

   C cos180°cos30° + sin180°sin30°

   D cos180°cos30° − sin180°sin30°

5. What is the exact value of cos75°?

   F $\frac{\sqrt{2} - \sqrt{6}}{2}$

   G $\frac{\sqrt{6} + \sqrt{2}}{4}$

   H $\frac{\sqrt{6} - \sqrt{2}}{4}$

   J $\frac{\sqrt{6} + \sqrt{2}}{2}$

Name ______________________ Date __________ Class __________

LESSON 14-4

Reading Strategy

Follow a Procedure

You can use sum and difference identities, along with the Unit Circle, to simplify expressions and find solutions.

Sine Sum Identity	Sine Difference Identity
$\sin(A + B) = \sin A \cos B + \cos A \sin B$	$\sin(A - B) = \sin A \cos B - \cos A \sin B$

To find the value of $\sin 105°$:

STEP 1: Determine whether the angle is the sum of two known angles, or the difference:
$105° = 45° + 60°$

STEP 2: Substitute the new angles into the sum or difference identity:
$\sin(105°) = \sin 45° \cos 60° + \cos 45° \sin 60°$

STEP 3: Solve:
$\sin(105°) = \sin 45° \cos 60° + \cos 45° \sin 60°$

$$= \frac{\sqrt{2}}{2} \bullet \frac{1}{2} + \frac{\sqrt{2}}{2} \bullet \frac{\sqrt{3}}{2}$$

$$= \frac{\sqrt{2}}{4} + \frac{\sqrt{6}}{4} = \frac{\sqrt{2} + \sqrt{6}}{4}$$

Identify a pair of angles whose trigonometric values are known and whose sum or difference is equal to each angle given.

1. 75° ______________

2. 255° ______________

3. 165° ______________

4. 105° ______________

Write the following as sum or difference identities.

5. $\sin(45° - 30°)$

____

6. $\sin(240° + 60°)$

____

7. $\sin(300° - 210°)$

____

8. $\sin(180° + 45°)$

____

9. $\sin(90° - 60°)$

____

10. $\sin(225° - 135°)$

____

Name ______________________ Date ____________ Class ____________

LESSON 14-5

Practice A
Double-Angle and Half-Angle Identities

Find sin 2θ, cos 2θ, and tan 2θ for each set of conditions.

1. $\sin\theta = -\frac{24}{25}$ for $\pi < \theta < \frac{3\pi}{2}$

a. Find $\cos\theta$ by substituting into formulas for $\sin 2\theta$ and $\cos 2\theta$. ____________

b. Use the double-angle identity $\sin 2\theta = 2\sin\theta\cos\theta$ to evaluate $\sin 2\theta$. ____________

c. Use the double-angle identity $\cos 2\theta = \cos^2\theta - \sin^2\theta$ to evaluate $\cos 2\theta$. ____________

d. Find $\tan\theta$ for substituting into formula for $\tan 2\theta$. ____________

e. Use $\tan 2\theta = \frac{2\tan\theta}{1 - \tan^2\theta}$ to evaluate $\tan 2\theta$. ____________

2. $\cos\theta = -\frac{3}{5}$ for $\frac{\pi}{2} < \theta < \pi$

____

3. $\sin\theta = \frac{5}{8}$ for $0 < \theta < \frac{\pi}{2}$

____

4. $\sin\theta = -\frac{1}{6}$ for $\frac{3\pi}{2} < \theta < 2\pi$

____

5. $\tan\theta = \frac{1}{3}$ for $\pi < \theta < \frac{3\pi}{2}$

____

Prove each identity.

6. $\sin 2\theta + 1 = (\sin\theta + \cos\theta)^2$

a. Expand the right side of the equation. ____________

b. Use the Associative Property of Addition. ____________

c. Use the double-angle identity for $\sin 2\theta$. ____________

d. Use the Pythagorean Identity. ____________

7. $\frac{\sin 2\theta}{2\cos^2\theta} = \tan\theta$

____

8. $2\cos^2\theta - 1 = \cos 2\theta$

____

Name ______________________ Date ____________ Class ____________

LESSON 14-5

Practice B

Double-Angle and Half-Angle Identities

Find sin 2θ, cos 2θ, and tan 2θ for each.

1. $\cos\theta = -\frac{12}{13}$ for $\pi < \theta < \frac{3\pi}{2}$

2. $\sin\theta = \frac{\sqrt{6}}{10}$ for $0 < \theta < \frac{\pi}{2}$

3. $\sin\theta = -\frac{2}{3}$ for $\frac{3\pi}{2} < \theta < 2\pi$

4. $\tan\theta = -\frac{5}{6}$ for $\frac{\pi}{2} < \theta < \pi$

Prove each identity.

5. $2\cos^2\theta = \cos 2\theta + 1$

6. $\tan\theta = \frac{1 - \cos 2\theta}{\sin 2\theta}$

Use half-angle identities to find the exact value of each trigonometric expression.

7. $\tan 22.5°$

8. $\cos\frac{7\pi}{12}$

9. $\sin\frac{11\pi}{12}$

Find $\sin\frac{\theta}{2}$, $\cos\frac{\theta}{2}$, and $\tan\frac{\theta}{2}$ for each.

10. $\cos\theta = \frac{3}{5}$ and $270° < \theta < 360°$

11. $\sin\theta = -\frac{\sqrt{5}}{3}$ and $180° < \theta < 270°$

Solve.

12. A water-park slide covers 100 feet of horizontal space and is 36 feet high.

 a. Write a trigonometric relation in terms of θ, the angle that the slide makes with the water surface.

 b. A new replacement slide will create an angle with the water surface that measures twice that of the original slide. The new slide will use the same horizontal space as the old slide. Write an expression that can be evaluated to find the height of the new slide.

 c. What is the height of the new slide to the nearest foot?

Name ______________________ Date ____________ Class ____________

LESSON 14-5

Practice C

Double-Angle and Half-Angle Identities

Find sin 2θ, cos 2θ, and tan 2θ for each.

1. $\cos\theta = \frac{\sqrt{3}}{10}$ for $\frac{3\pi}{2} < \theta < 2\pi$

2. $\tan\theta = \frac{4}{7}$ for $\pi < \theta < \frac{3\pi}{2}$

3. $\cos\theta = -\frac{\sqrt{5}}{6}$ for $\frac{\pi}{2} < \theta < \pi$

4. $\sin\theta = \frac{3\sqrt{6}}{16}$ for $0 < \theta < \frac{\pi}{2}$

Prove each identity.

5. $\cos 2\theta = \frac{\sin 2\theta - \tan\theta}{\tan\theta}$

6. $\sin 4\theta = 4\cos\theta\sin\theta - 8\cos\theta\sin^3\theta$

Use half-angle identities to find the exact value of each trigonometric expression.

7. $\sin 157.5°$

8. $\cos\frac{5\pi}{12}$

9. $\tan\frac{13\pi}{8}$

Find $\sin\frac{\theta}{2}$, $\cos\frac{\theta}{2}$, and $\tan\frac{\theta}{2}$ for each.

10. $\sin\theta = \frac{1}{5}$ for $90° < \theta < 180°$

11. $\cos\theta = \frac{\sqrt{7}}{4}$ for $270° < \theta < 360°$

12. $\tan\theta = 7$ for $180° < \theta < 270°$

13. $\tan\theta = -\frac{\sqrt{5}}{2}$ for $270° < \theta < 360°$

Solve.

14. Amanda wants to shorten the wheelchair ramp at her front door. The current ramp is 13 feet along the horizontal and the step is 9 inches high. If she doubles the measure of the angle that the ramp makes with the walkway, how long will the new ramp be to the nearest inch?

Name ______________________ Date __________ Class __________

LESSON 14-5

Reteach

Double-Angle and Half-Angle Identities

There are special double-angle formulas for $\sin 2\theta$, $\cos 2\theta$, and $\tan 2\theta$.

The double angle is twice θ, or 2θ.

Double-Angle Identities
$\sin 2\theta = 2\sin\theta\cos\theta$
$\cos 2\theta = \cos^2\theta - \sin^2\theta$
$\cos 2\theta = 2\cos^2\theta - 1$
$\cos 2\theta = 1 - 2\sin^2\theta$
$\tan 2\theta = \dfrac{2\tan\theta}{1 - \tan^2\theta}$

There are three equivalent forms of the identity for $\cos 2\theta$. If you know one form, you can use $\sin^2\theta + \cos^2\theta = 1$ to derive the other forms.

Find $\sin 2\theta$ and $\cos 2\theta$ if $\cos\theta = -\frac{2}{3}$ and $180° < \theta < 270°$.

Notice the interval in which θ lies. Cosine and sine are both negative in Quadrant III.

Step 1 Find $\sin\theta$.

$\sin^2\theta = 1 - \cos^2\theta$ *Pythagorean identity*

$\sin^2\theta = 1 - \left(-\frac{2}{3}\right)^2$ *Substitute $\cos\theta = -\frac{2}{3}$.*

$\sin^2\theta = \frac{5}{9}$ *Evaluate.*

$\sin\theta = -\frac{\sqrt{5}}{3}$ *Solve for $\sin\theta$.*

Use the negative value since θ lies in Quadrant III.

Step 2 Find $\sin 2\theta$.

$\sin 2\theta = 2\sin\theta\cos\theta$ *Write the formula.*

$\sin 2\theta = 2\left(-\frac{\sqrt{5}}{3}\right)\left(-\frac{2}{3}\right)$ *Substitute values for $\sin\theta$ and $\cos\theta$.*

$\sin 2\theta = \frac{4\sqrt{5}}{9}$ *Simplify.*

Step 3 Find $\cos 2\theta$.

$\cos 2\theta = 2\cos^2\theta - 1$ *Choose a formula.*

$\cos 2\theta = 2\left(-\frac{2}{3}\right)^2 - 1$ *Substitute the value for $\cos\theta$.*

$\cos 2\theta = \frac{8}{9} - 1 = -\frac{1}{9}$ *Simplify.*

Evaluate each expression if $\cos\theta = \frac{1}{4}$ and $270° < \theta < 360°$.

1. $\sin\theta$
   Use $\sin^2\theta = 1 - \cos^2\theta$.

   ____

2. $\cos 2\theta$
   Use $\cos 2\theta = 1 - 2\sin^2\theta$.

   ____

3. $\tan\theta$
   Use $\tan\theta = \dfrac{\sin\theta}{\cos\theta}$.

   ____

4. $\tan 2\theta$
   Use $\tan 2\theta = \dfrac{2\tan\theta}{1 - \tan^2\theta}$.

   ____

Name ______________________ Date __________ Class __________

LESSON 14-5

Reteach

Double-Angle and Half-Angle Identities (continued)

You can use half-angle identities to find the exact value of some trigonometric expressions.

Half-Angle Identities
$\sin\left(\frac{\theta}{2}\right) = \pm\sqrt{\frac{1 - \cos\theta}{2}}$: (+) for $\frac{\theta}{2}$ in Quadrants I, II; (−) for $\frac{\theta}{2}$ in Quadrants III, IV
$\cos\left(\frac{\theta}{2}\right) = \pm\sqrt{\frac{1 + \cos\theta}{2}}$: (+) for $\frac{\theta}{2}$ in Quadrants I, IV; (−) for $\frac{\theta}{2}$ in Quadrants II, III
$\tan\left(\frac{\theta}{2}\right) = \pm\sqrt{\frac{1 - \cos\theta}{1 + \cos\theta}}$: (+) for $\frac{\theta}{2}$ in Quadrants I, III; (−) for $\frac{\theta}{2}$ in Quadrants II, IV

Find the exact value of cos 112.5°.

$\cos 112.5° = \cos\frac{225°}{2}$ — *Find the angle:* $\theta = 225°$.

Think: 225° has a reference angle of 45°.

$= -\sqrt{\frac{1 + \cos\theta}{2}}$ — *Choose the half-angle identity.*

112.5° is in QII where cosine is negative.

$= -\sqrt{\frac{1 + \cos 225°}{2}}$ — *Substitute* $\theta = 225°$.

$= -\sqrt{\frac{1 + \left(-\frac{\sqrt{2}}{2}\right)}{2}}$ — *Evaluate.*

$\cos 225° = -\cos 45° = -\frac{\sqrt{2}}{2}$

$= -\sqrt{\left(\frac{2 - \sqrt{2}}{2}\right)\left(\frac{1}{2}\right)}$ — *Simplify.*

$= -\sqrt{\frac{2 - \sqrt{2}}{4}} = -\frac{\sqrt{2 - \sqrt{2}}}{2}$

Use the half-angle identity to find the exact value of each expression.

5. $\tan 157.5° = \tan\frac{315°}{2}$

$\tan\left(\frac{315°}{2}\right) = -\sqrt{\frac{1 - \cos 315°}{1 + \cos 315°}}$

____

____

6. $\sin 15° = \sin\frac{30°}{2}$

$\sin\left(\frac{30°}{2}\right) =$ ______________________

____

____

Name ______________________ Date __________ Class __________

LESSON 14-5

Challenge

Double-Angle and Half-Angle Identities for Tangent

Double-angle and half-angle identities for sine and cosine are shown below.

Double-Angle Identities	*Half-Angle Identities*
$\sin 2A = 2\sin A \cos A$	$\sin \frac{A}{2} = \pm\sqrt{\frac{1 - \cos A}{2}}$
$\cos 2A = \cos^2 A - \sin^2 A$	$\cos \frac{A}{2} = \pm\sqrt{\frac{1 + \cos A}{2}}$

You can use these identities to derive double-angle and half-angle identities for tangent.

1. **a.** Use a ratio identity to express $\tan 2A$ in terms of $\sin 2A$ and $\cos 2A$. ____________

   b. Use identities for $\sin 2A$ and $\cos 2A$ to rewrite the right side in terms of $\sin A$ and $\cos A$. ____________

   c. Divide numerator and denominator of the right side by $\cos^2 A$. ____________

   d. Simplify, and rewrite the right side in terms of $\tan A$. ____________

2. Verify the result of Exercise 1 by using the identity shown below to obtain an identity for $\tan 2A$.
   $\tan(A + B) = \frac{\tan A + \tan B}{1 - \tan A \tan B}$ ____________

3. Angle A is in Quadrant II and $\sin A = \frac{4}{5}$.

   a. Use your identity to determine $\tan 2A$. ____________

   b. Use a calculator to verify your result. ____________

4. Show that $2\cot 2A = \cot A - \tan A$ is an identity by

   a. expressing $2\cot 2A$ in terms of $\tan A$. ____________

   b. expressing $\cot A - \tan A$ as a single fraction in terms of $\tan A$. ____________

5. Derive an identity for $\tan\frac{A}{2}$ in terms of $\cos A$. ____________

6. **a.** Use your identity to determine $\tan 22.5°$. ____________

   b. Verify that your result is equivalent to $\sqrt{2} - 1$.

   ____

7. Show that $\tan \frac{A}{2}$ is equivalent to $\frac{1 - \cos A}{\sin A}$.

   ____

Name ______________________ Date __________ Class __________

LESSON 14-5

Problem Solving

Double-Angle and Half-Angle Identities

Chris set a new personal best in football with a 35-yard kick. He wants to improve his distance. His coach suggests using the formula $d(\theta) = \frac{3v_0^2 \sin\theta \cos\theta}{16}$, which gives the horizontal distance, *d*, in yards, as a function of the angle of elevation, θ, at which the ball is kicked with an initial velocity of v_0 yards per second.

1. What is the angle of elevation of the ball when Chris kicks it with an initial velocity of 20 yards per second and the ball covers 35 yards.
   a. What double-angle identity can you use to rewrite the function in terms of $\sin 2\theta$? ______________
   b. Rewrite the function in terms of $\sin 2\theta$. ______________
   c. Substitute the known values in the equation. ______________
   d. Solve for $\sin 2\theta$. ______________
   e. What is the angle at which he kicked the ball? ______________
2. Chris wonders how the horizontal distance would change if he kicked the ball with the same velocity but at a different angle.
   a. Complete the table to show the horizontal distance, to the nearest hundredth yard, for the given angles.
   b. At what angle must Chris kick the ball to cover the greatest horizontal distance?

   ____

Angle of Elevation (°)	Horizontal Distance (yd)
15	
25	
35	
45	
55	

3. What is the least angle at which Chris must kick the ball if he wants to cover at least 30 yards horizontally?

   ____
4. Chris thinks he can kick the ball with an initial velocity of 18 yards per second and make a horizontal distance of 40 yards. Is he correct? Explain.

   ____

Choose the letter for the best answer.

5. If Chris kicks the ball with an initial velocity of only 15 yards per second and an angle of elevation of 45°, what is the maximum horizontal distance the ball will cover?

   A 12.06 yd **C** 21.09 yd
   B 14.92 yd **D** 29.83 yd
6. At what angle can Chris kick the ball to cover a horizontal distance of 36 yards, if he kicks the ball with an initial velocity of 20 yards per second?

   F 18.5° **H** 44.1°
   G 36.9° **J** 73.7°

Name ______________________ Date __________ Class __________

LESSON 14-5

Reading Strategy
Analyze Information

You can use double-angle identities and half-angle identities to evaluate expressions and find exact values. One step is to find the sign of the functions of the angle.

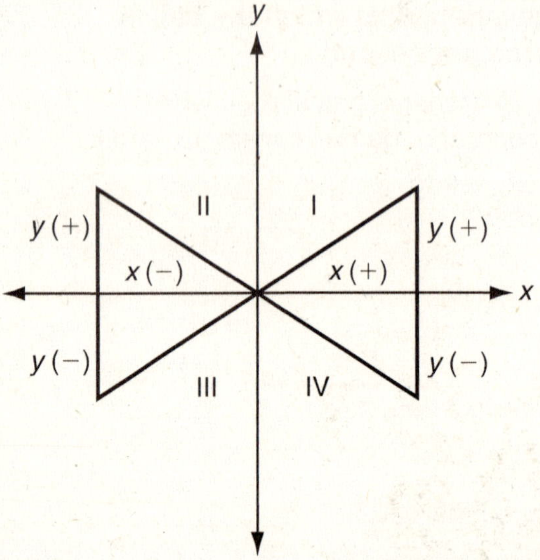

Use the diagram above to determine the sign of each function of θ in each quadrant. Write + or − next to the function name.

	Quadrant I	Quadrant II	Quadrant III	Quadrant IV
1.	sin	sin	sin	sin
2.	cos	cos	cos	cos
3.	tan	tan	tan	tan

Determine the quadrant and sign of each trigonometric expression.

4. sin330° ______________

5. cos210° ______________

6. sin150° ______________

7. tan123° ______________

8. sin290° ______________

9. cos13° ______________

10. sin194° ______________

11. cos359° ______________

12. tan275° ______________

13. cos178° ______________

14. tan205° ______________

15. sin85° ______________

Name ______________________ Date __________ Class __________

LESSON 14-6

Practice A

Solving Trigonometric Equations

Find all of the solutions of each equation.

1. $3\sin\theta = \sin\theta - \sqrt{3}$

a. Subtract $\sin\theta$ from both sides and combine like terms. ______________________

b. Solve for $\sin\theta$. ______________________

c. Apply the inverse sine. ______________________

2. $5\cos\theta = 3\cos\theta - 1$ ______________________

3. $10\tan\theta = 9\tan\theta + 1$ ______________________

Solve each equation for the given domain.

4. $\cos^2\theta - 2\cos\theta = -1$ for $0° \le \theta \le 360°$

a. Add 1 to both sides. ______________________

b. Factor or use the quadratic formula to solve for $\cos\theta$. ______________________

c. Apply the inverse cosine. ______________________

5. $\tan^2\theta = 9$ for $0° \le \theta \le 360°$ ______________________

6. $4\sin^2\theta = 3\sin\theta$ for $0° \le \theta \le 360°$ ______________________

Solve each equation for the given domain. Use trigonometric identities.

7. $\cos^2\theta = \sin^2\theta + \sin\theta$ for $0° \le \theta \le 360°$

a. Rewrite the equation in terms of $\sin\theta$. ______________________

b. Solve for $\sin\theta$. ______________________

c. Apply the inverse sine. ______________________

8. $\cos 2\theta + \sin\theta = 1$ for $0° \le \theta \le 360°$ ______________________

9. $\sin 2\theta \tan\theta = 1$ for $0° \le \theta \le 360°$ ______________________

Solve.

10. The horizontal range, *R*, in feet that a baseball travels is modeled by $R(\theta) = \frac{v^2 \sin 2\theta}{32}$. The initial velocity, *v*, is 90 feet per second. The ball is pitched at an angle of θ degrees with the horizontal. At what angle(s) will the ball travel 200 feet? ______________________

Name ______________________ Date ____________ Class ____________

LESSON 14-6

Practice B

Solving Trigonometric Equations

Find all of the solutions of each equation.

1. $4\tan\theta = 5\tan\theta + \sqrt{3}$

____

2. $\sqrt{2} - 2\sin\theta = 0$

____

3. $7\cos\theta - 1 = 9\cos\theta$

____

4. $2\tan\theta + \sqrt{3} = 5\tan\theta$

____

Solve each equation for the given domain.

5. $\tan^2\theta - 2\tan\theta = -1$ for $0° \le \theta \le 360°$

____

6. $2\sin^2\theta = 1$ for $0° \le \theta \le 360°$

____

7. $4\cos^2\theta = 3\cos\theta$ for $90° \le \theta \le 180°$

____

8. $2\sqrt{3}\cos^2\theta - \cos\theta = 2\sqrt{3}$ for $\pi \le \theta \le 2\pi$

____

9. $3\tan^2\theta = 2\tan\theta + 1$ for $0° \le \theta \le 90°$

____

10. $5\tan^2\theta = 2\tan\theta + 6$ for $180° \le \theta \le 360°$

____

Use trigonometric identities to solve each equation for the given domain.

11. $\sin\theta + \sin 2\theta = 0$ for $0° \le \theta \le 360°$

____

12. $\cos^2\theta = \sin^2\theta$ for $0° \le \theta \le 360°$

____

13. $\dfrac{\cos 2\theta}{\sin^2\theta} = 0$ for $0° \le \theta \le 360°$

____

14. $\cos 2\theta = \frac{1}{2}\sin\theta$ for $0° \le \theta \le 90°$

____

Solve.

15. The height of the water at a pier on a certain day can be modeled by $h(t) = 4.8\sin\frac{\pi}{6}(t + 3.5) + 9$, where h is the height in feet and t is the time in hours after midnight. When is the height of the water 6 feet?

____

Name ______________________ Date ____________ Class ____________

LESSON 14-6

Practice C

Solving Trigonometric Equations

Find all of the solutions of each equation.

1. $4\tan\theta + \sqrt{3} = \tan\theta$

2. $\cos\theta = 1 + 3\cos\theta$

Solve each equation for the given domain.

3. $4\cos^2\theta = 1$ for $0° \le \theta \le 360°$

4. $\cos^2\theta + \cos\theta - 1 = 0$ for $0° \le \theta \le 180°$

5. $3\tan^2\theta + \sqrt{3} = 3\tan\theta + \sqrt{3}\tan\theta$ for $\pi \le \theta \le 2\pi$

Use trigonometric identities to solve each equation for the given domain.

6. $\sin 2\theta = \sqrt{2}\cos\theta$ for $0° \le \theta \le 360°$

7. $\cos\frac{\theta}{2} = \sin\theta$ for $0° \le \theta \le 360°$

8. $\tan 2\theta = \cot\theta$ for $0° \le \theta \le 180°$

9. $2\cos^2\frac{\theta}{2} = \cos 2\theta$ for $0° \le \theta \le 360°$

10. $\sin\left(\frac{\pi}{4} - \theta\right) = 1 - \sin\left(\frac{\pi}{4} + \theta\right)$ for $0 \le \theta \le 2\pi$

11. $\cos 2\theta \tan\theta = -\sin\theta$ for $0° \le \theta \le 360°$

12. $\cos\theta + \sin\theta = 1$ for $0° \le \theta \le 360°$

13. $\cos^3\theta - \sin^3\theta = 1$ for $0° \le \theta \le 360°$

Solve.

14. The number of hours of daylight, *d*, at Quintilla Falls can be modeled by $h(d) = 3.22\sin\frac{\pi}{182.5}(d - 88) + 12.32$, where *d* is the number of days after January 1. On what dates are there 12 hours of sunlight? Assume the year is not a leap year.

Name ______________________ Date ____________ Class ____________

LESSON 14-6

Reteach

Solving Trigonometric Equations

You can use the same methods to solve trigonometric equations as to solve algebraic equations.

Substitute variables to solve trigonometric equations that resemble quadratic equations.

Solve: $\sin^2\theta - 3\sin\theta - 4 = 0$ for $0° \le \theta < 360°$

> The interval includes 0° but not 360°.

Step 1 Make a substitution.

Let $x = \sin\theta$ in $\sin^2\theta - 3\sin\theta - 4 = 0$

$$\downarrow \quad \downarrow \quad \downarrow \quad \downarrow$$

$$x^2 \quad - 3x \quad -4 = 0$$

Step 2 Solve the quadratic $x^2 - 3x - 4 = 0$.

$(x - 4)(x + 1) = 0$		*Factor.*
$x - 4 = 0$	$x + 1 = 0$	*Set each factor equal to 0.*
$x = 4$	$x = -1$	*Solve each equation for x.*

Step 3 Substitute $\sin\theta$ for x.

> $-1 \le \sin\theta \le 1$

$\sin\theta = 4$ — *This equation has no solution.*

$\sin\theta = -1$ — *Think: What angle has a sine of −1?*

$\theta = 270°$

The only solution is $\theta = 270°$.

Complete to solve each equation for $0° \le \theta < 360°$.

1. $3\tan^2\theta - 1 = 0$

Let $x = \tan\theta$.

Hint: There are 4 solutions.

$3x^2 - 1 = 0$

$3x^2 = 1$

$x = \pm\dfrac{\sqrt{3}}{3}$

____

____

2. $2\sin^2\theta - \sin\theta = 0$

Let $x = \sin\theta$.

Hint: There are 4 solutions.

____

____

____

____

____

3. $\cos^2\theta - 2\cos\theta + 1 = 0$

____

____

4. $2\cos^2\theta - 7\cos\theta + 3 = 0$

____

____

Name ______________________ Date __________ Class __________

LESSON 14-6

Reteach

Solving Trigonometric Equations (continued)

You can use trigonometric identities to solve trigonometric equations. This is especially useful when more than one trigonometric function appears in an equation.

> Use an identity to rewrite the equation using a single trigonometric function or a single angle.

Solve: $\cos 2\theta + \cos\theta = 0$ for $0° \leq \theta < 360°$

> Use an identity. Choose the formula that allows you to write the equation in terms of $\cos\theta$ only.

Step 1 The angles are not the same.
Use the double-angle identity for cosine.
Choose $\cos 2\theta = 2\cos^2\theta - 1$.

Step 2 Substitute $2\cos^2\theta - 1$ for $\cos 2\theta$ in the original equation.

$\cos 2\theta + \cos\theta = 0$ — *Write the original equation.*

$(2\cos^2\theta - 1) + \cos\theta = 0$ — *Substitute.*

$2\cos^2\theta + \cos\theta - 1 = 0$ — *Write in standard form.*

Step 3 Solve the equation.

$2\cos^2\theta + \cos\theta - 1 = 0$

$(2\cos\theta - 1)(\cos\theta + 1) = 0$ — *Factor.*

$2\cos\theta - 1 = 0$ $\cos\theta + 1 = 0$ — *Set each factor equal to 0.*

$\cos\theta = \frac{1}{2}$ $\cos\theta = -1$ — *Solve each equation for $\cos\theta$.*

Step 4 Find the values that satisfy each solution.

$\cos\theta = \frac{1}{2}$ — Think: What angle has a cosine of $\frac{1}{2}$?

$\theta = 60°, 300°$

$\cos\theta = -1$ — Think: What angle has a cosine of -1?

$\theta = 180°$

> Remember to find all values in the interval.

Use trigonometric identities to solve each equation for $0° \leq \theta < 360°$.

5. $\cos 2\theta + \sin\theta + 2 = 0$

Use $\cos 2\theta = 1 - 2\sin^2\theta$.

____

6. $\cos^2\theta + 2\sin\theta = 1$

Use $\cos^2\theta = 1 - \sin^2\theta$.

____

Name ______________________ Date __________ Class __________

LESSON 14-6

Challenge

Trigonometric Equations with Radicals

A *trigonometric equation* contains one or more trigonometric functions of an angle. Such an equation must first be solved for a function of the angle and then for the measure of the angle. Some trigonometric equations contain radicals.

1. Solve $2\cos\theta = \sqrt{8\sin\theta - 1}$ for $0° \le \theta < 360°$.

   a. Since the radical is isolated, square both sides of the equation. ______________

   b. Use an identity to express the equation in terms of a single trigonometric function. ______________

   c. Solve for θ and check. ______________

Solve each equation for $0° \le \theta < 360°$. Check your results.

2. $\sqrt{\dfrac{\sin\theta}{2}} = \sin\theta$

____

3. $\sqrt{2 + 3\cot^2\theta} = \tan\theta$

____

4. $\sin\theta + \sin\dfrac{\theta}{2} = 0$

   a. Use an identity to replace $\sin\dfrac{\theta}{2}$. ______________

   b. Isolate the radical and square both sides. ______________

   c. Use an identity to express the equation in terms of a single trigonometric function. ______________

   d. Solve for θ and check. ______________

5. $\cos\theta = \sin\dfrac{\theta}{2}$

____

6. $\sin\theta = 2\sin\dfrac{\theta}{2}$

____

As opposed to the equations in Exercises 4–6, where you first had to get the angles to be of the same multiple of θ, the angles in Exercises 7 and 8 are already of the same multiple. So, you do not need to use radicals.

Solve each equation for $0° \le \theta < 360°$.

7. $\cos\dfrac{\theta}{2} + \sin\dfrac{\theta}{2} = 1$

____

8. $2\sin\theta\cos\theta + 2\sin\theta - \cos\theta - 1 = 0$

____

Name ______________________ Date ___________ Class ___________

LESSON 14-6

Problem Solving
Solving Trigonometric Equations

Jon watches his sister Jessica skateboard. When she jumps and launches herself into the air with an initial speed of v_0 feet per second, her path in terms of time, t, in seconds, is represented by these equations.

Equation 1 $x(t) = v_0 t\cos\theta$

Equation 2 $y(t) = v_0 t\sin\theta - 16t^2$

The first equation models the horizontal distance, $x(t)$, that the skateboarder travels, and the second equation models the vertical height, $y(t)$, that the skateboarder attains.

1. Jessica attains a height of 4.7 feet above the launch and landing ramps after 1 second. Her initial velocity is 25 feet per second. Find the angle of her launch.

   a. Which equation can you use with the given information to solve for θ? ____________

   b. Substitute the known values and solve for θ. ____________

   c. What is Jessica's height above the launch and landing ramps after 0.5 second? ____________

   d. What distance has Jessica traveled after 1 second? ____________

2. Jon attains a height of 5.2 feet above the launch and landing ramps after 1 second. His initial velocity is 28 feet per second. Find the angle of his launch.

   a. Write and evaluate an expression for Jon's launch angle.

   ____

   b. Write and evaluate an expression for his height above the launch and landing ramps after 0.5 second.

   ____

Choose the letter for the best answer.

3. How far does Jon travel in 1 second?

   A 14.5 ft
   C 18.3 ft
   B 16.7 ft
   D 21.2 ft

4. Jessica increases her initial velocity to 30 feet per second. She attains a height of 5.5 feet after 1 second. Which expression represents her launch angle?

   F $\cos^{-1}\left(\frac{10.5}{30}\right)$
   H $\sin^{-1}\left(\frac{10.5}{30}\right)$
   G $\cos^{-1}\left(\frac{21.5}{30}\right)$
   J $\sin^{-1}\left(\frac{21.5}{30}\right)$

Name ______________________ Date ____________ Class ____________

LESSON 14-6

Reading Strategy
Compare and Contrast

You can solve trigonometric equations in the same way as you solve algebraic equations.

Linear Algebraic Equations	Linear Trigonometric Equations
$3x = x + 2$ $3x - x = 2$ $2x = 2$ $x = 1$	$3\tan\theta = \tan\theta + 2$ $3\tan\theta - \tan\theta = 2$ $2\tan\theta = 2$ $\tan\theta = 1$
Quadratic Algebraic Equations	**Quadratic Trigonometric Equations**
$x^2 + 3x - 10 = 0$ $(x + 5)(x - 2) = 0$ $x + 5 = 0$ or $x - 2 = 0$ $x = -5$ or $x = 2$	$2\cos^2\theta - \cos\theta - 1 = 0$ $(2\cos\theta + 1)(\cos\theta - 1) = 0$ $2\cos\theta + 1 = 0$ or $\cos\theta - 1 = 0$ $\cos\theta = -\frac{1}{2}$ or $\cos\theta = 1$

Solve.

1. $3\cot\theta = \cot\theta + 1$

____

2. $2\sec\theta = 6\sec\theta - 3$

____

3. $\frac{5}{2}\cos\theta + \frac{3}{2}\cos\theta - 4 = 0$

____

4. $5\tan\theta + 3 = 4\tan\theta - \frac{1}{3}$

____

5. $6\sin\theta = 3\sin\theta + 3$

____

6. $\frac{1}{2}\cot\theta = -\frac{5}{2}\cot\theta + 2$

____

7. $3\cos^2\theta - 1 = 2\cos^2\theta - \frac{3}{4}$

____

8. $\sin^2\theta + 3\sin\theta + 5 = \sin\theta + 4 + \sin^2\theta$

____

9. $3\tan^2\theta = \tan\theta$

____

10. $4\cos^2\theta = 1 - 3\cos\theta$

____

LESSON 14-1

Practice A

Graphs of Sine and Cosine

Identify whether each function is periodic. If the function is periodic, give the period.

1. 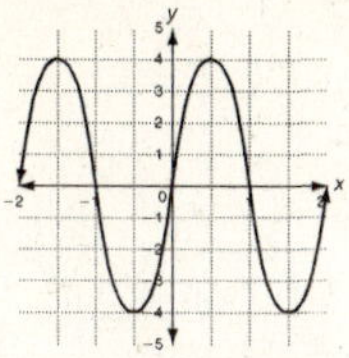

Periodic; 2

2. 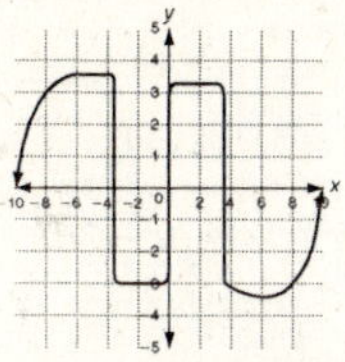

Not periodic

Use $f(x) = \sin x$ or $g(x) = \cos x$ as a guide. Identify the amplitude and period. Then graph each function.

3. $h(x) = -3\sin(4x)$

Amplitude: 3; period: $\frac{\pi}{2}$

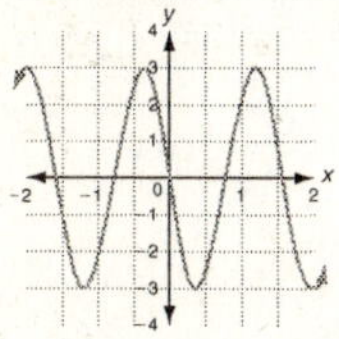

4. $p(x) = 2\cos(0.5x)$

Amplitude: 2; period: 4π

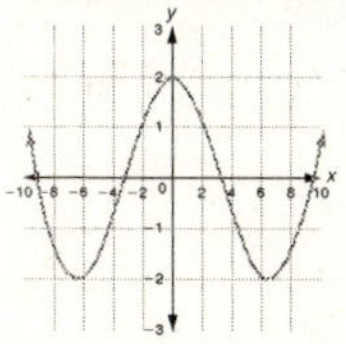

Use $f(x) = \cos x$ as a guide. Graph the function.

5. $g(x) = \cos(x - \pi)$

a. Identify the amplitude and period.

Amplitude: 1; period: 2π

b. Identify the phase shift.

π radians to the right

c. Identify the x-intercepts.

$\frac{\pi}{2} + n\pi$, where n is an integer

d. Identify the maximum and minimum values.

1, −1

e. Use the information to graph the function.

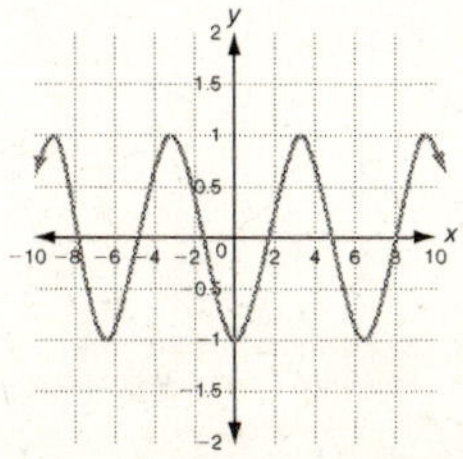

LESSON 14-1

Practice B

Graphs of Sine and Cosine

Using $f(x) = \sin x$ or $g(x) = \cos x$ as a guide, graph each function. Identify the amplitude and period.

1. $b(x) = -5\sin \pi x$

Amplitude: 5; period: 2

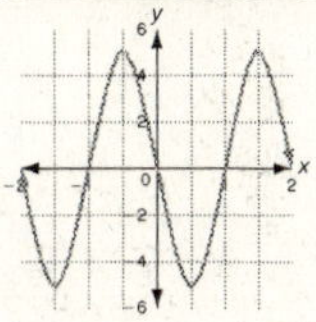

2. $k(x) = 3\cos 2\pi x$

Amplitude: 3; period: 1

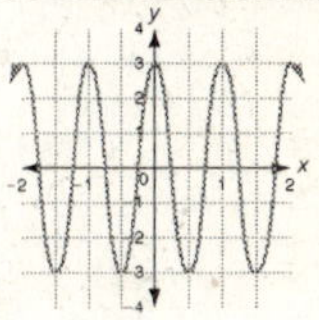

Using $f(x) = \sin x$ or $g(x) = \cos x$ as a guide, graph each function. Identify the x-intercepts and phase shift.

3. $h(x) = \sin\left(x + \frac{\pi}{4}\right)$

x-intercepts: $\frac{3\pi}{4}, \frac{7\pi}{4}$; phase shift: $\frac{\pi}{4}$ radians to the left

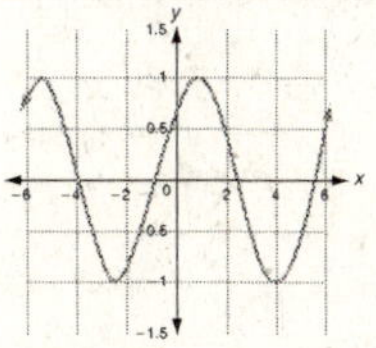

4. $h(x) = \cos\left(x - \frac{\pi}{4}\right)$

x-intercepts: $\frac{3\pi}{4}, \frac{7\pi}{4}$; phase shift: $\frac{\pi}{4}$ radians to the right

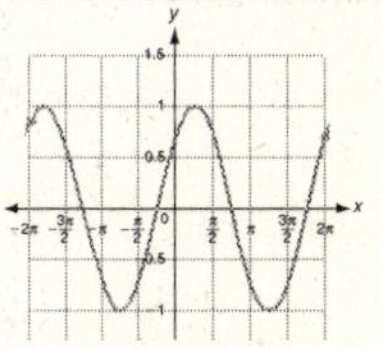

Solve.

5. a. Use a sine function to graph a sound wave with a period of 0.002 second and an amplitude of 2 centimeters.

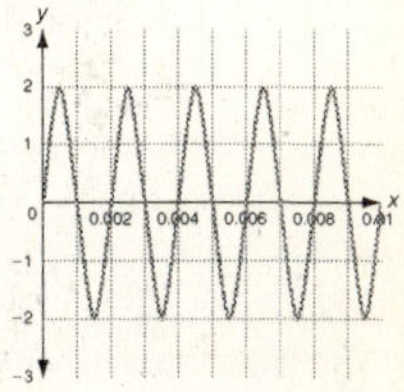

b. Find the frequency in hertz for this sound wave.

500 Hz

LESSON 14-1

Practice C

Graphs of Sine and Cosine

Using $f(x) = \sin x$ or $f(x) = \cos x$ as a guide, graph each function. Identify the amplitude, period, x-intercepts, and phase shift.

1. $h(x) = \frac{1}{2}\cos(-\pi x)$

Amplitude: 0.5; period: 2; x-intercepts: 0.5, 1.5; phase shift: 0

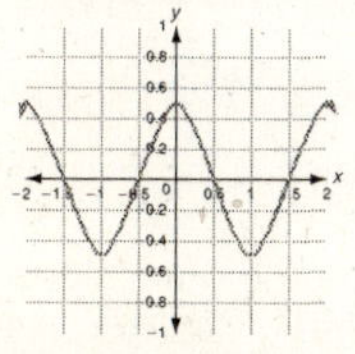

2. $q(x) = -\sin\left(\frac{\pi}{2}x\right)$

Amplitude: 1; period: 4; x-intercepts: 0, 2, 4; phase shift: 0

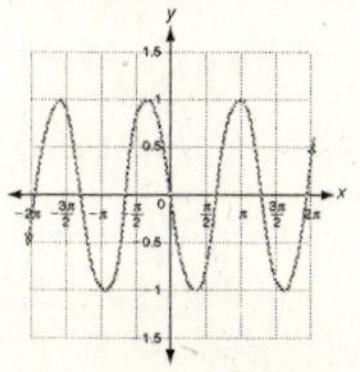

3. $c(x) = 3\cos(x + \pi)$

Amplitude: 3; period: 2π; x-intercepts: $\frac{\pi}{2}, \frac{3\pi}{2}$; phase shift: π radians to the left

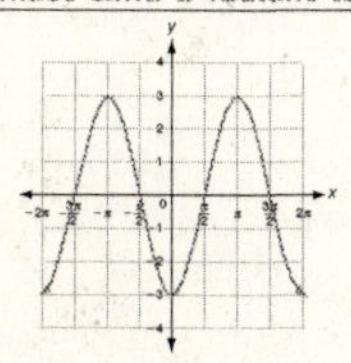

4. $h(x) = -2\sin\left(x - \frac{\pi}{2}\right)$

Amplitude: 2; period: 2π; x-intercepts: $\frac{\pi}{2}, \frac{3\pi}{2}$; phase shift: $\frac{\pi}{2}$ radians to the right

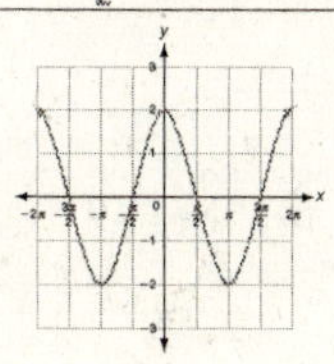

Solve.

5. A manual metronome is an inverted pendulum that helps musicians play to the beat. The number of centimeters, C, that the tip of the pendulum is from a tabletop can be modeled by $C(t) = 2\cos 4\pi t + 12$, where t is the time in seconds.

a. Graph the height of the pendulum tip for 2 periods.

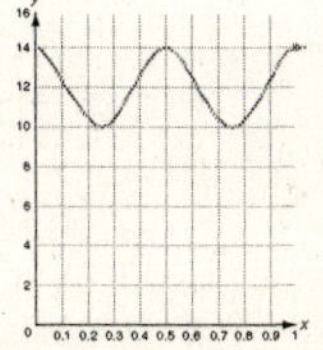

b. How high is the pendulum when $t = \frac{1}{4}$ second?

10 cm

LESSON 14-1

Reteach

Graphs of Sine and Cosine

Transformations of the sine and cosine functions change the amplitude and/or the period of the graph.

For $y = a\sin bx$ or $y = a\cos bx$:

- the amplitude is $|a|$, — The amplitude is half the difference between the greatest and least values of the function.
- the period is $\frac{2\pi}{|b|}$. — One full cycle appears in each period.

Use the graph of $f(x) = \sin x$ to sketch the graph of $g(x) = 0.5\sin 2x$.

Step 1 Compare $g(x) = 0.5\sin 2x$ to $y = a\sin bx$. Find a to identify the amplitude.

$a = 0.5$ and $|0.5| = 0.5$, so the amplitude is 0.5. — The maximum value of $g(x)$ is 0.5 and the minimum value is −0.5.

Step 2 Find b to identify the period.

$b = 2$, and $\frac{2\pi}{|b|} = \frac{2\pi}{|2|} = \pi$, so the period is π. — One full cycle appears in the interval from 0 to π.

Step 3 Graph $f(x) = \sin x$.

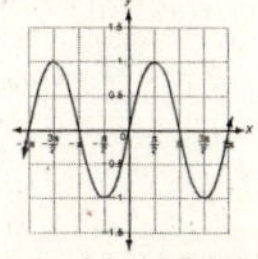

The **amplitude** is 1. The maximum and minimum values of $f(x)$ are 1 and −1. The **period** is 2π. One full cycle appears in the interval from 0 to 2π. Two full cycles appear in the interval from -2π to 2π. The x-intercepts are at multiples of π.

Step 4 Graph $g(x) = 0.5\sin 2x$ on the same plane as $f(x)$.

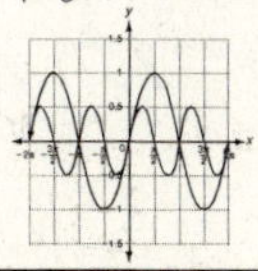

The **amplitude** is 0.5. The maximum and minimum values of $g(x)$ are 0.5 and −0.5. The **period** is π. One full cycle appears in the interval from 0 to π. Two full cycles appear in the interval from 0 to 2π and from -2π to 0. The x-intercepts are at multiples of $\frac{\pi}{2}$.

Complete to graph $h(x) = 0.5\cos 2x$.

1. Find the amplitude of $h(x)$. $a =$ 0.5
2. Find the period of $h(x)$. $\frac{2\pi}{|b|} =$ π
3. What are the maximum and minimum values of $h(x)$? 0.5, −0.5
4. How many full cycles appear in the interval from 0 to π? 1 cycle
5. Sketch the graph of $f(x) = \cos x$. Then graph $h(x) = 0.5\cos 2x$.

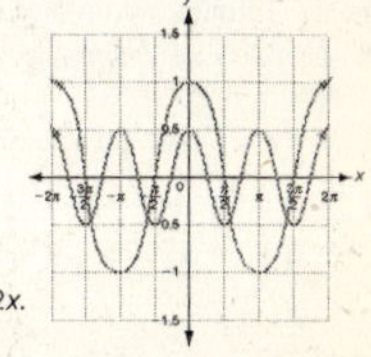

LESSON 14-1

Reteach

Graphs of Sine and Cosine (continued)

A **phase shift** is a horizontal translation. Sine and cosine can be translated horizontally by $y = \sin(x - h)$ and $y = \cos(x - h)$.

A phase shift (or horizontal translation) of h units moves the graph left h units for $h < 0$ or right h units for $h > 0$.

Use the graph of $f(x) = \cos x$ to sketch the graph of $g(x) = \cos\left(x - \frac{\pi}{2}\right)$.

Step 1 Compare $g(x) = \cos\left(x - \frac{\pi}{2}\right)$ to $y = \boldsymbol{a}\cos \boldsymbol{b}x$.

Find the amplitude and period.

$a = 1$ and $|1| = 1$, so the amplitude is 1.

$b = 1$, and $\frac{2\pi}{|b|} = \frac{2\pi}{|1|} = 2\pi$, so the period is 2π.

> The amplitude and period of g are the same as for $y = \cos x$.

Step 2 Find h to identify the phase shift.

$x - h = x - \frac{\pi}{2}$, so $h = \frac{\pi}{2}$.

The phase shift is $\frac{\pi}{2}$ radians to the right.

> Because $h > 0$, the shift is to the right.

Step 3 Identify the first two positive x-intercepts.

The x-intercepts of $f(x) = \cos x$ occur at $\frac{\pi}{2}$ and $\frac{3\pi}{2}$.

The x-intercepts of $g(x)$ occur at $\frac{\pi}{2} + \frac{\pi}{2}$, or π, and $\frac{3\pi}{2} + \frac{\pi}{2}$, or 2π.

> Intercepts occur at integer multiples of π.

Step 4 Identify the maximum and minimum values.

The maxima and minima of $f(x) = \cos x$ occur at 0 and π.

The maxima and minima of $g(x)$ occur at $0 + \frac{\pi}{2}$, or $\frac{\pi}{2}$, and $\pi + \frac{\pi}{2}$, or $\frac{3\pi}{2}$.

Step 5 Graph $f(x) = \cos x$ and $g(x) = \cos\left(x - \frac{\pi}{2}\right)$.

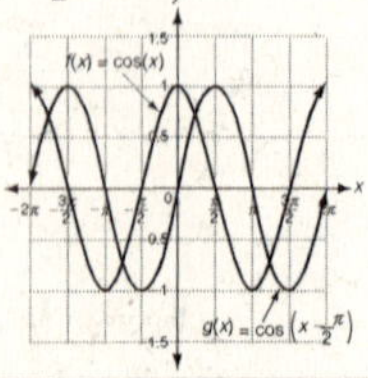

Use the graph of $f(x) = \sin x$ to sketch the graph of $g(x) = \sin\left(x - \frac{\pi}{2}\right)$.

6. Identify h. What is the phase shift?

   $h = \frac{\pi}{2}$; phase shift: $\frac{\pi}{2}$ radians to the right

7. Identify the x-intercepts from 0 to 2π.

   $\frac{\pi}{2}, \frac{3\pi}{2}$

8. Identify the maxima and minima from 0 to 2π.

   Maximum of 1 at π; minimum of −1 at 0 and 2π

9. Sketch the graphs of $f(x)$ and $g(x)$.

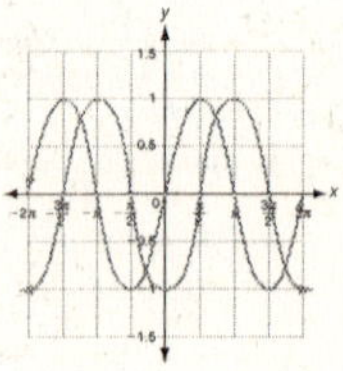

LESSON 14-1

Challenge

Sinusoidal Functions as Mathematical Models

In many real-world situations, a dependent variable repeats its values at regular intervals as the independent variable changes. In such cases, a sine or cosine function would be a reasonable mathematical model.

As represented at right, a waterwheel with a radius of 7 feet rotates at 6 revolutions per minute. You start a stopwatch when the point P on the rim of the wheel is at its lowest possible position. After 5 seconds, point P is at its greatest height.

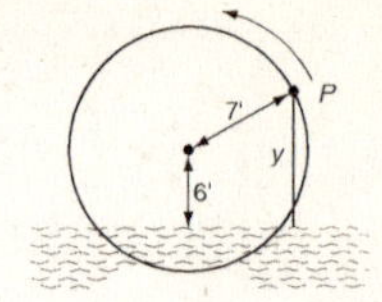

To write an equation that will model the distance, y, of point P from the surface of the water in terms of the number of seconds, x, on the stopwatch, assume that y varies sinusoidally with x, and begin with a sine equation.

1. In the general sine equation, $y = a \sin b(x - h) + k$, identify the constants that need to be determined. **a, b, h, k**
2. Since the center is 6 feet above the water, the horizontal line of symmetry of the sine curve will be 6 units above the x-axis. This information gives you the value of one of the constants. Write the value of this constant. **$k = 6$**
3. Since the radius of the wheel is 7 feet, point P goes 7 feet above and below the center of the wheel. From this information, write the value of another constant. **$a = 7$**
4. The wheel makes 6 complete revolutions every 60 seconds. From this information, write the value of another constant. **$b = \frac{\pi}{5}$**
5. P was at its maximum after 5 seconds. Use this information to write the value of the last constant. **$h = \frac{5}{2}$**
6. Write the required equation to model the movement of the waterwheel. **$y = 7\sin\frac{\pi}{5}\left(x - \frac{5}{2}\right) + 6$**
7. Sketch the graph of your equation on the coordinate plane below.

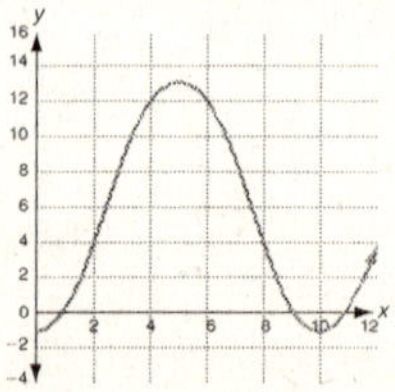

8. Use your graph to predict the height of point P above the surface of the water when the stopwatch shows 7 seconds.

   about 8.2 ft

LESSON 14-1

Problem Solving

Graphs of Sine and Cosine

As a result of the tide, the depth of the water in the bay varies with time. According to the harbormaster, the depth can be modeled by the function $d(t) = 2.5\cos\left(\frac{\pi}{6}\right)(t - 2) + 4$, where t is the number of hours since midnight, and d is the depth in meters.

1. Identify the following features of the graph of the function.
   a. Amplitude **2.5**
   b. Period **12**
   c. Phase shift **2 units to the right**
   d. Vertical shift **4 units up**
   e. What is the maximum depth and when does that occur? **6.5 m at 2:00 A.M.**
   f. What is the minimum depth and when does that occur? **1.5 m at 8:00 A.M.**

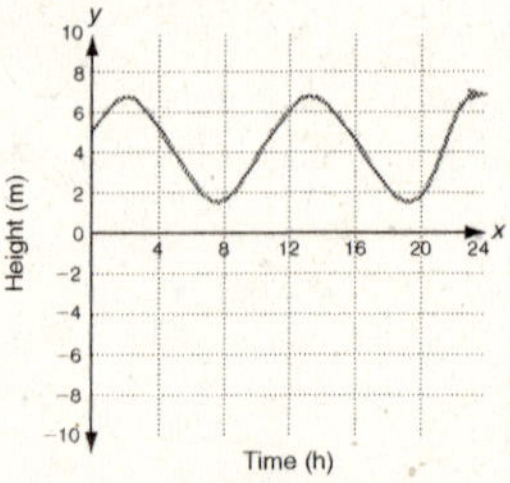

2. Graph the function for 24 hours.
3. How many high tides occur in 24 hours? **2**
4. Jim thinks that the amplitude and the vertical shift should be reversed.
   a. Write the equation for this new function. **$d(t) = 4\cos\left(\frac{\pi}{6}\right)(t - 2) + 2.5$**
   b. Explain how water depth and frequency of tides would change using Jim's function. **Possible answer: The maximum depth of the water would still be 6.5 m, but the minimum depth would be −1.5 m. The bay would dry out. The frequency of high tide and low tide would not change.**
5. Petra thinks that the original function is correct, except for the period, which should be 24. If so, how many hours after midnight would the first high and low tides occur? **2 h and 14 h**

Jim looks at equations for the depth of water in different locations. Choose the letter for the best answer.

6. Which function represents a location that will dry out?
   (A) $d(t) = 5\cos\frac{\pi}{12}(t - 1) + 4$
   B $d(t) = 4\cos\frac{\pi}{8}(t - 2) + 4.5$
   C $d(t) = 3\cos\frac{\pi}{6}(t - 3) + 5$
   D $d(t) = 2\cos\frac{\pi}{4}(t - 4) + 5.5$

7. How deep is the water at 1:00 A.M. in a location where the water depth t hours after midnight is modeled by $d(t) = 1.25\cos\frac{\pi}{12}(t - 1) + 6$?
   F 4.75 m
   G 5.75 m
   (H) 7.25 m
   J 8.25 m

LESSON 14-1

Reading Strategy

Use a Visual Map

The form of the sine or cosine function determines how the graph varies. Use the graph of $\sin x$ as a reference.

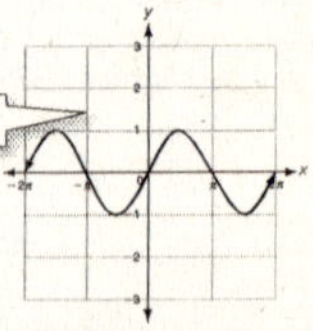

$y = a\sin b(x - h) + k$

The absolute value of a, $|a|$, tells you the value of the amplitude of the function.
$y = \mathbf{2}\sin x$

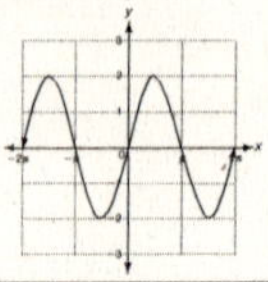

The value of the vertical shift, or vertical translation, is represented by k.
$y = \sin(x) + \mathbf{2}$

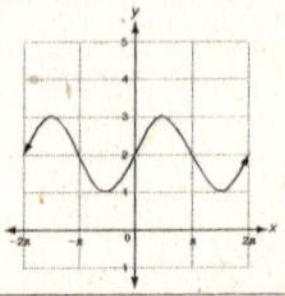

From the variable b you can find the period, $\frac{2\pi}{|b|}$. The period tells you the length of one complete cycle.
$y = \sin(\mathbf{2}x)$

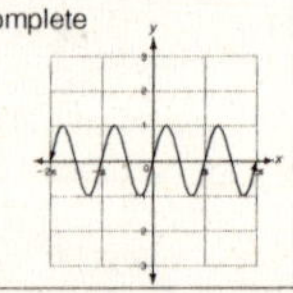

The value of the phase shift, or horizontal translation, is represented by h.
$y = \sin(x - \mathbf{2})$

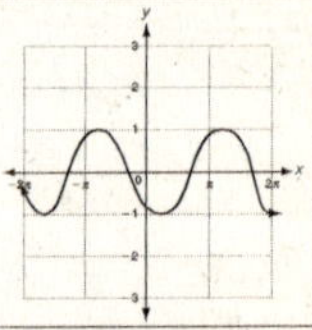

Answer each question.

1. Describe how the period of $f(x) = \sin x$ compares to the period of $g(x) = 4\sin 3x$.

   The period of $g(x)$, $\frac{2\pi}{3}$, is $\frac{1}{3}$ the size of the period of $f(x)$, 2π.

2. Describe how the period of $f(x) = \cos x$ compares to the period of $g(x) = \frac{1}{2}\cos(x - 3)$.

   The period of $g(x)$ is equal to the period of $f(x)$, 2π.

3. Describe the phase shift for the function $g(x) = 2\sin 3(x - 8) + 5$.

   The phase shift is 8; the function is translated 8 units to the right.

4. What are the amplitude and period for the function $g(x) = 3\sin\frac{1}{2}(x + 2) - 1$?

   The amplitude is 3, and the period is 4π.

5. What are the phase shift and the vertical shift for the function $g(x) = \cos 4(x + 5) + 3$?

   The phase shift is −5 or 5 units left, and the vertical shift is 3 units up.

LESSON 14-2

Practice A
Graphs of Other Trigonometric Functions

Graph each function. Use $f(x) = \tan x$ as a guide. Identify the period, x-intercepts, and asymptotes.

1. $p(x) = \tan(3x)$
   a. Identify the period. $\frac{\pi}{3}$
   b. Identify the x-intercepts. $\frac{\pi}{3}n$
   c. Identify the asymptotes. $\frac{\pi}{6} + \frac{\pi n}{3}$
   d. Use the information to graph the function.

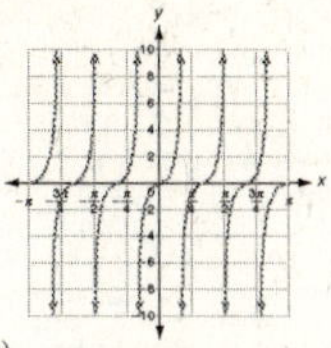

2. $q(x) = \tan\left(\frac{x}{2}\right)$
   Period: 2π; x-intercepts: $2\pi n$; asymptotes: $\pi + 2\pi n$

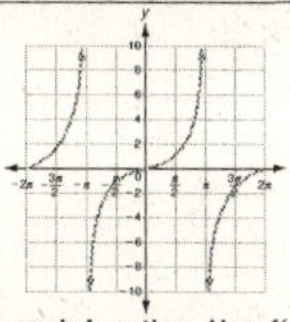

3. $k(x) = \tan(\pi x)$
   Period: 1; x-intercepts: n; asymptotes: $\frac{1}{2} + n$

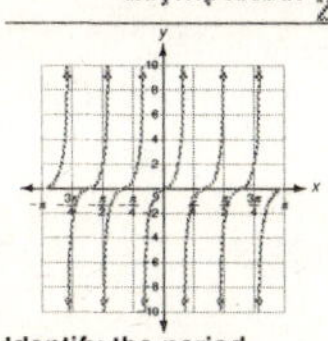

Graph each function. Use $f(x) = \cot x$ as a guide. Identify the period, x-intercepts, and asymptotes.

4. $h(x) = 2\cot\left(\frac{x}{2}\right)$
   a. Identify the period. 2π
   b. Identify the x-intercepts. $\pi + 2\pi n$
   c. Identify the asymptotes. $2\pi n$
   d. Use the information to graph the function.

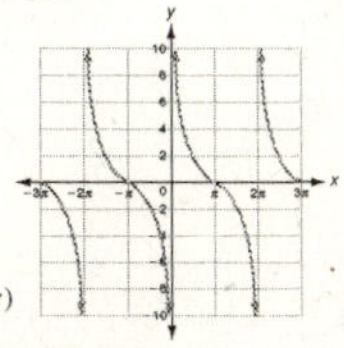

5. $b(x) = \frac{1}{2}\cot(2x)$
   Period: $\frac{\pi}{2}$; x-intercepts: $\frac{\pi}{4} + \frac{\pi n}{2}$; asymptotes: $\frac{\pi n}{2}$

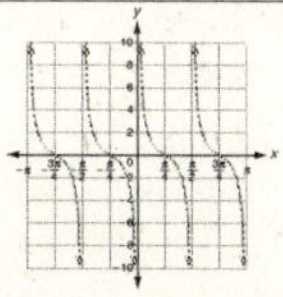

6. $b(x) = \frac{1}{2}\cot(x)$
   Period: π; x-intercepts: $\frac{\pi}{2} + \pi n$; asymptotes: πn

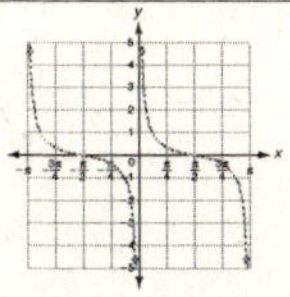

LESSON 14-2

Practice B
Graphs of Other Trigonometric Functions

Using $f(x) = \tan x$ and $f(x) = \cot x$ as a guide, graph each function. Identify the period, x-intercepts, and asymptotes.

1. $g(x) = 2\tan\frac{\pi x}{2}$
   Period: 2; x-intercepts: $2n$; asymptotes: $1 + 2n$

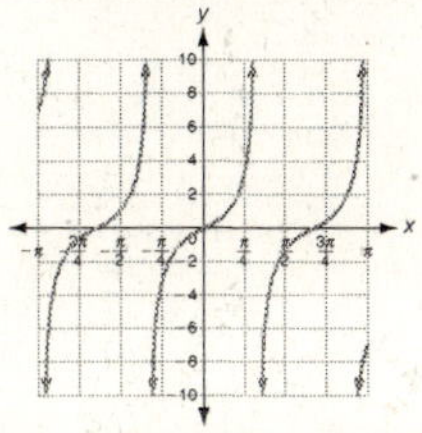

2. $t(x) = \frac{3}{4}\cot(x)$
   Period: π; x-intercepts: $\frac{\pi}{2} + \pi n$; asymptotes: πn

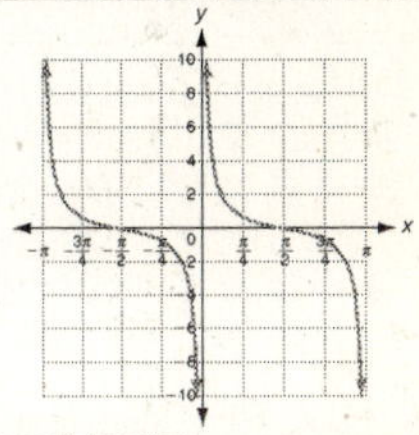

Using $f(x) = \cos x$ or $f(x) = \sin x$ as a guide, graph each function. Identify the period and asymptotes.

3. $k(x) = \sec\frac{x}{2}$
   Period: 4π; asymptotes: $\pi + 2\pi n$

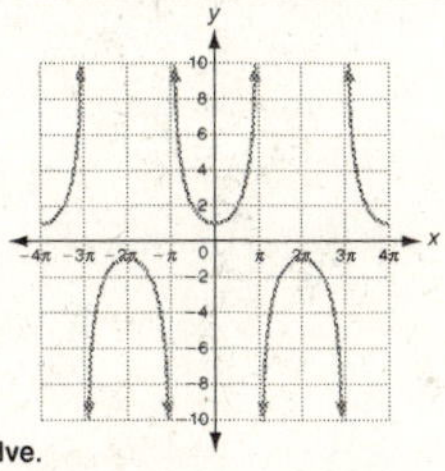

4. $q(x) = \frac{1}{2}\csc(2x)$
   Period: π; asymptotes: $\frac{\pi n}{2}$

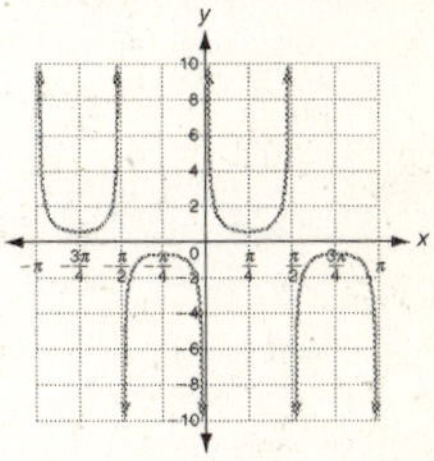

Solve.

5. The rotating light on a lighthouse is 400 feet from a cliff and completes one full rotation every 10 seconds. The equation representing the distance, a, in feet that the center of the circle of light is from the lighthouse is $a(t) = 400\sec\left(\frac{\pi t}{5}\right)$.
   a. What is the period of $a(t)$? 10
   b. Find the value of the function at $t = 10$. 400 ft

LESSON 14-2

Practice C
Graphs of Other Trigonometric Functions

Using $f(x) = \tan x$ or $f(x) = \cot x$ as a guide, graph each function. Identify the period, x-intercepts, and asymptotes.

1. $k(x) = \frac{1}{4}\tan(\pi x)$
   Period: 1; x-intercepts: n; asymptotes: $\frac{1}{2} + n$

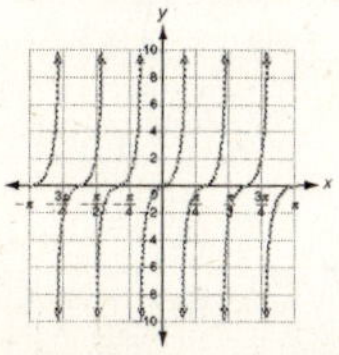

2. $c(x) = \cot\left(\frac{2x}{3}\right)$
   Period: $\frac{3\pi}{2}$; x-intercepts: $\frac{3\pi}{4} + \frac{3\pi n}{2}$; asymptotes: $\frac{3\pi n}{2}$

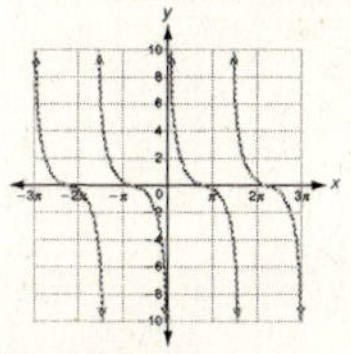

Using $f(x) = \cos x$ or $f(x) = \sin x$ as a guide, graph each function. Identify the period and asymptotes.

3. $k(x) = \sec\frac{x}{4}$
   Period: 8π; asymptotes: $2\pi + 4\pi n$

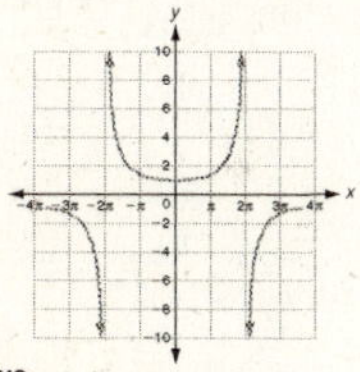

4. $q(x) = 2\csc(2x)$
   Period: π; asymptotes: $\frac{\pi n}{2}$

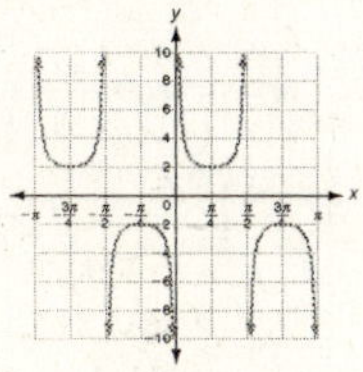

Solve.

5. A strobe light is located in the center of a square dance room. The rotating light is 40 feet from each of the 4 walls and completes one full rotation every 6 seconds. The equation representing the distance, d, in feet that the center of the circle of light is from the light source is $d(t) = \left|40\sec\left(\frac{\pi t}{3}\right)\right|$.
   a. What is the period of $d(t)$? 6
   b. Find the value of the function at $t = 2.5$. 46.2 ft

LESSON 14-2

Reteach
Graphs of Other Trigonometric Functions

Transformations of the tangent function change the period and/or asymptotes of the graph.

For $y = a\tan bx$: (Note: x is measured in radians.)

- the period is $\frac{\pi}{|b|}$,
- the asymptotes are located at $x = \frac{\pi}{2|b|} + \frac{\pi n}{|b|}$, where n is an integer.

Use the graph of $f(x) = \tan x$ to sketch the graph of $g(x) = \tan\frac{1}{2}x$.

Step 1 Find b to identify the period.
$b = \frac{1}{2}$, and $\frac{\pi}{|b|} = \frac{\pi}{\left|\frac{1}{2}\right|} = 2\pi$, so the period is 2π.
(There will be one full cycle between $-\pi$ and π.)

Step 2 Use the period to identify the x-intercepts.
The first x-intercepts of both $f(x)$ and $g(x)$ occur at 0. Because the period is 2π, the intercepts occur at $2\pi n$. For example, -2π, 0, and 2π are x-intercepts.
(The intercepts shift 2π radians right and occur at integer multiples of 2π or the even multiples of π.)

Step 3 Identify the asymptotes.
$b = \frac{1}{2}$, so the asymptotes occur at $x = \frac{\pi}{2\left|\frac{1}{2}\right|} + \frac{\pi n}{\left|\frac{1}{2}\right|} = \pi + 2\pi n$.

Step 4 Graph $f(x) = \tan x$ and $g(x) = \tan\frac{1}{2}x$ on the same plane.

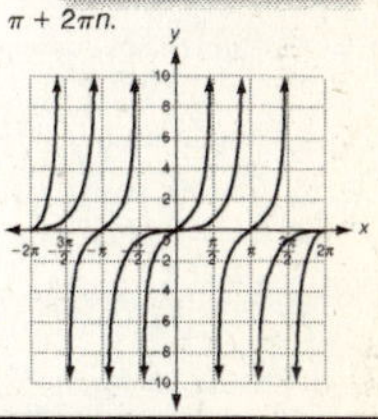

The x-intercepts of $g(x)$ are also x-intercepts of $f(x)$. The asymptotes are different. Because the period of $g(x)$ is 2π, there are fewer cycles of $g(x)$ in the same interval as $f(x)$.

Complete to graph $g(x) = \tan 2x$. Use the interval from $-\pi$ to π.

1. Find the period of $g(x)$. $\frac{\pi}{|b|} =$ $\frac{\pi}{2}$
2. Find the x-intercepts of $g(x)$.
   $-\pi, -\frac{\pi}{2}, 0, \frac{\pi}{2}, \pi$
3. Find the asymptotes of $g(x)$.
   $x = -\frac{3\pi}{4}, -\frac{\pi}{4}, \frac{\pi}{4}, \frac{3\pi}{4}$
4. Sketch the graph of $f(x) = \tan x$. Then graph $g(x) = \tan 2x$.

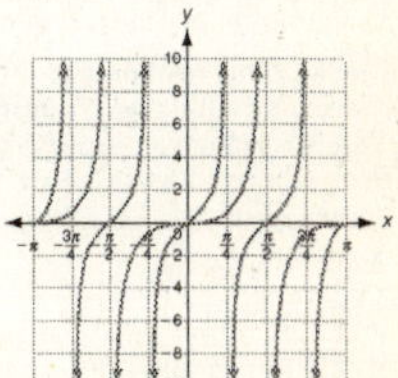

LESSON 14-2

Reteach

Graphs of Other Trigonometric Functions (continued)

Transformations of the cotangent function are similar to transformations of the tangent function.

For $y = a\cot bx$:

- the period is $\frac{\pi}{|b|}$. — This is the same as for tangent transformations.
- the asymptotes are located at $x = \frac{\pi n}{|b|}$, where n is an integer.

Use the graph of $f(x) = \cot x$ to sketch the graph of $g(x) = \cot 2x$.

Step 1 Find b to identify the period.

$b = 2$, and $\frac{\pi}{|b|} = \frac{\pi}{|2|} = \frac{\pi}{2}$, so the period is $\frac{\pi}{2}$. — There will be one full cycle between 0 and $\frac{\pi}{2}$.

Step 2 Use the period to identify the x-intercepts.

The first x-intercept of $f(x)$ is $\frac{\pi}{2}$.

The first x-intercept of $g(x)$ occurs at $\frac{\pi}{2} \cdot \frac{1}{2}$, or $\frac{\pi}{4}$. — The x-intercepts shift $\frac{\pi}{2}$ radians right.

From $-\pi$ to π the x-intercepts are $-\frac{3\pi}{4}, -\frac{\pi}{4}, \frac{\pi}{4}$, and $\frac{3\pi}{4}$.

Step 3 Identify the asymptotes.

$b = 2$, so the asymptotes occur at $x = \frac{\pi n}{|2|} = \frac{\pi n}{2}$. — These are the multiples of $\frac{\pi}{2}$.

Step 4 Graph $g(x) = \cot 2x$.

- Sketch the asymptotes.
- Plot the intercepts.
- Then use $f(x) = \cot x$ as a guide to sketch the graph of $g(x) = \cot 2x$.

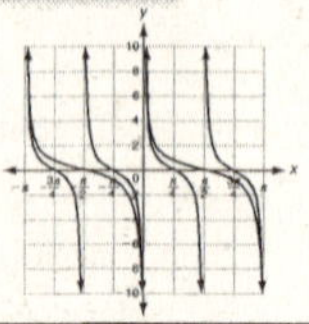

Complete to graph $g(x) = \cot\frac{1}{2}x$. Use the interval from -2π to 2π.

5. Find the period of $g(x)$. $\frac{\pi}{|b|} =$ 2π
6. Find the x-intercepts of $g(x)$. $-\pi, \pi$
7. Where are the asymptotes of $g(x)$? $x = -2\pi, 0, 2\pi$
8. Sketch the graph of $g(x) = \cot\frac{1}{2}x$.

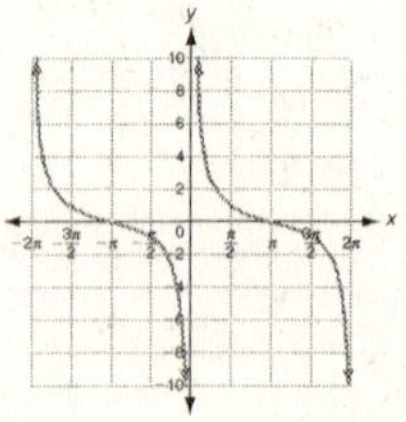

LESSON 14-2

Challenge

Trigonometric Functions from the Graph

Consider the graph at right. It looks like a tangent graph with some transformations. The graph of $y = \tan x$ has the center point of one period pass through the origin. The center of the middle period of this graph passes through the point $(1, -2)$. This indicates that both a vertical translation and a phase shift are included in the equation of the function. The period is 2π and indicates a stretch in the period of the graph of the parent function, $y = \tan x$ by a factor of 2. Using these observations, an equation for the function is $y = \tan(0.5(x - 1)) - 2$. Due to the periodic nature of the tangent function, other equations are possible.

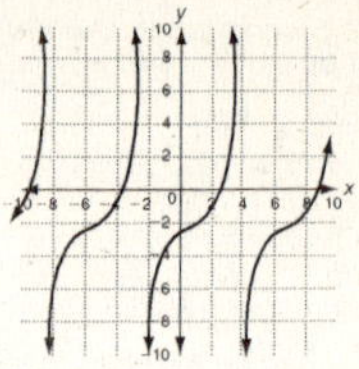

Write an equation for each graph below. Possible Answers:

1. 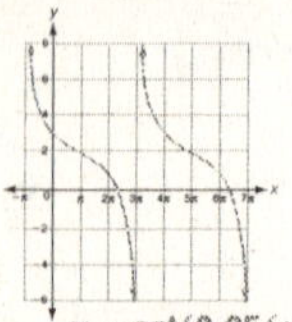
$y = \cot(0.25(x + 3)) + 2$

2. 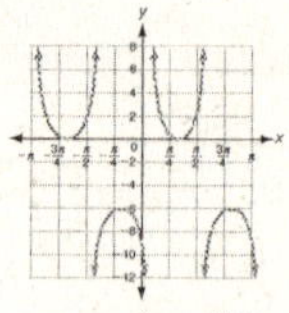
$y = 3\sec(2(x - 1)) - 3$

3. 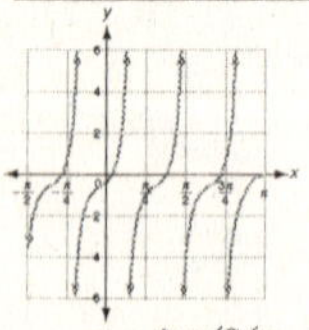
$y = \tan(3(x - 2)) - 1$

4. 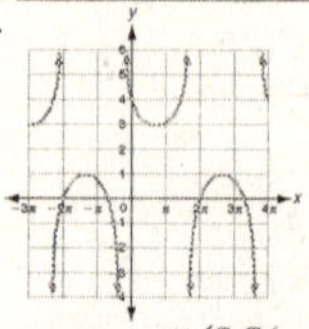
$y = \csc(0.5(x + 1)) + 2$

5. A helicopter flying in a straight line at a constant altitude of 2000 feet will pass directly over a playground. Let y be the distance from the playground to a point on the ground directly beneath the helicopter as shown in the diagram. Let x be the angle of elevation from a person on the playground to the helicopter. Write y as a function of x and sketch a graph over the interval from 0 to π. $y = 2000\cot x$

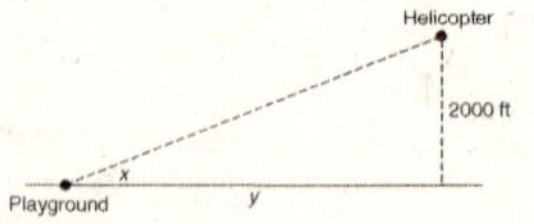

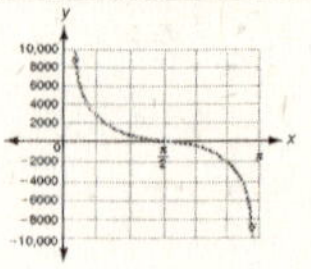

LESSON 14-2

Problem Solving

Graphs of Other Trigonometric Functions

As the sign outside the community youth center rotates, the light on it traces a path on the center's wall. The distance in feet of the light to the point on the wall where it shines varies with time, t, and can be modeled by the function $g(t) = 9\csc\left(\frac{3\pi}{4}t\right)$, where t is measured in seconds.

1. Identify the period of the function.
   a. What is the coefficient of t in the equation of the function? $\frac{3\pi}{4}$
   b. Write and evaluate an expression for the period of the function. $\frac{2\pi}{\left(\frac{3\pi}{4}\right)} = \frac{8}{3}$
   c. Explain what the period of the function means in terms of the time for one rotation of the light.
   Possible answer: It takes $2\frac{2}{3}$ s for the light to make one full rotation.
   d. How does the graph of the function vary based on the period?
   Possible answer: The graph repeats every $2\frac{2}{3}$ seconds.
2. a. This function has asymptotes at points where what other function is equal to 0?
   $\sin\left(\frac{3\pi}{4}t\right)$
   b. What are the asymptotes of the function?
   $t = \frac{4}{3}n$
3. What is the significance of 9 in the equation for the function?
   Possible answers: It is the shortest distance from the light to the point on the wall where it shines; or it is the value of $g(t)$ when $\csc\left(\frac{3\pi}{4}t\right) = 1$.
4. Graph the function using all the information about the function.

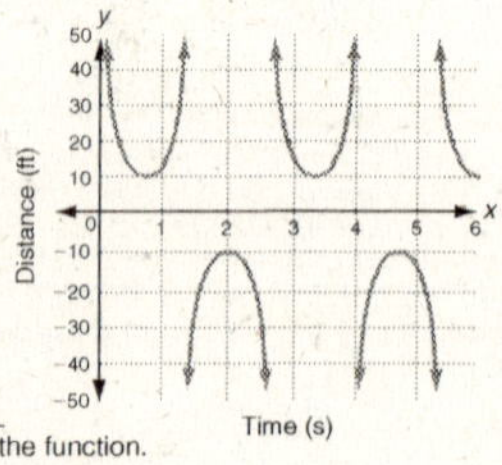

Toni changes the rotation rate of the light so that it rotates once every 4 seconds. Choose the letter for the best answer.

5. Which equation represents the distance of the light to the point on the wall where it shines, as a function of time, t?
   A $g(t) = 9\csc\left(\frac{\pi}{4}t\right)$
   (B) $g(t) = 9\csc\left(\frac{\pi}{2}t\right)$
   C $g(t) = 9\csc(2\pi t)$
   D $g(t) = 9\csc(4\pi t)$
6. Which pair of parameters describes this new function?
   (F) period: 4 s; vertical asymptotes at $\pm 2n$ s
   G period: 2 s; vertical asymptotes at $\pm n$ s
   H period: 2 s; horizontal asymptotes at $\pm 2n$ s
   J period: 4 s; horizontal asymptotes at $\pm n$ s

LESSON 14-2

Reading Strategy

Use Graphic Aids

Tangent and cotangent graphs have different characteristics.

In **tangent graphs** of the form $y = a\tan bx$, the period is $\frac{\pi}{\lvert b\rvert}$ and the asymptotes are at $x = \frac{\pi}{2\lvert b\rvert} + \frac{\pi n}{\lvert b\rvert}$. $y = \tan x$ 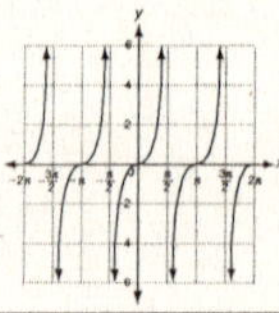	In **cotangent graphs** of the form $y = a\cot bx$, the period is $\frac{\pi}{\lvert b\rvert}$ and the asymptotes are at $x = \frac{\pi n}{\lvert b\rvert}$. $y = \cot x$ 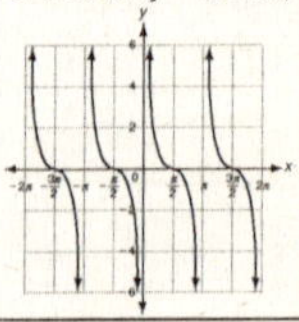

Determine whether each graph is a tangent or cotangent function.

1. 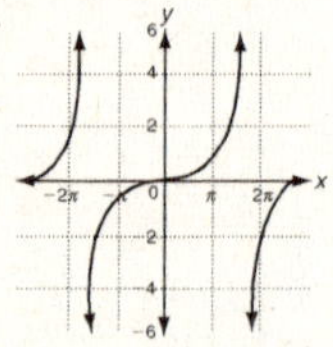
Tangent

2. 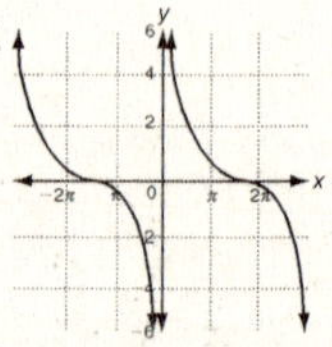
Cotangent

Identify the period and asymptotes of each graph or function.

3. 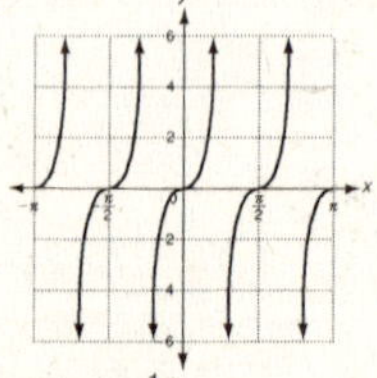
   a. Identify the period. $\frac{\pi}{2}$
   b. Identify the asymptotes. $x = \frac{\pi}{4} + \frac{\pi n}{2}$
4. $g(x) = \tan\frac{1}{2}x$
   a. Identify the period. 2π
   b. Identify the asymptotes. $x = \pi + 2\pi n$
5. $g(x) = \cot 2x$
   a. Identify the period. $\frac{\pi}{2}$
   b. Identify the asymptotes. $x = \frac{\pi n}{2}$

LESSON 14-3

Practice A

Fundamental Trigonometric Identities

Prove each trigonometric identity.

1. $\frac{\sin\theta}{1-\cos^2\theta} = \csc\theta$

   a. Modify the left-hand side. Replace 1 with a known identify. $\frac{\sin\theta}{\cos^2\theta + \sin^2\theta - \cos^2\theta}$

   b. Simplify the denominator. $\frac{\sin\theta}{\sin^2\theta}$

   c. Keep simplifying and substituting identities until the left side matches the right side. $\frac{1}{\sin\theta} = \csc\theta;\ \csc\theta = \csc\theta$

2. $\cos^4\theta + \cos^2\theta\sin^2\theta + \sin^2\theta = 1$

   $\cos^4\theta + \cos^2\theta\sin^2\theta + \sin^2\theta = 1$
   $\cos^2\theta(\cos^2\theta + \sin^2\theta) + \sin^2\theta = 1$
   $\cos^2\theta(\cos^2\theta + \sin^2\theta) + \sin^2\theta = 1$
   $\cos^2\theta(1) + \sin^2\theta = 1$
   $\cos^2\theta + \sin^2\theta = 1$
   $1 = 1$

3. $\cot\theta = \frac{\csc\theta}{\sec\theta}$

   $\cot\theta = \frac{\csc\theta}{\sec\theta};\ \cot\theta = \frac{\frac{1}{\sin\theta}}{\frac{1}{\cos\theta}};$
   $\cot\theta = \frac{\cos\theta}{\sin\theta};\ \cot\theta = \cot\theta$

Rewrite each expression in terms of cos θ. Then simplify.

4. $\frac{\cot\theta}{2\csc\theta}$

   a. Rewrite each function in terms of sin θ or cos θ. $\frac{\frac{\cos\theta}{\sin\theta}}{\frac{2}{\sin\theta}}$

   b. Rewrite the fraction in simplest form. $\frac{\sin\theta\cos\theta}{2\sin\theta} = \frac{\cos\theta}{2}$

5. $\sin\theta\cot\theta$

   $\sin\theta\cot\theta =$
   $\sin\theta \cdot \frac{\cos\theta}{\sin\theta} =$
   $\cos\theta$

6. $\frac{\cos\theta - \cos^2\theta}{\sin^2\theta}$

   $\frac{\cos\theta - \cos^2\theta}{\sin^2\theta} =$
   $\frac{\cos\theta - \cos^2\theta}{1-\cos^2\theta} =$
   $\frac{\cos\theta(1-\cos\theta)}{(1+\cos\theta)(1-\cos\theta)} =$
   $\frac{\cos\theta}{1+\cos\theta}$

7. $\frac{\sin^2\theta + 2\cos^2\theta}{\sec^2\theta - \tan^2\theta}$

   $\frac{\sin^2\theta + 2\cos^2\theta}{\sec^2\theta - \tan^2\theta} =$
   $\frac{1-\cos^2\theta + 2\cos^2\theta}{\frac{1}{\cos^2\theta} - (\sec^2\theta - 1)} =$
   $\frac{1+\cos^2\theta}{\frac{1}{\cos^2\theta} - \frac{1}{\cos^2\theta} + 1} =$
   $1 + \cos^2\theta$

Solve.

8. Use the equation $mg\sin\theta = \mu mg\cos\theta$ to determine the angle at which a desk can be tilted before a paperback book on the desk begins to slide. Assume $\mu = 1.11$. 48°

LESSON 14-3

Practice B

Fundamental Trigonometric Identities

Prove each trigonometric identity.

1. $\sin^2\theta + \sin^2\theta\cot^2\theta = 1$

   $\sin^2\theta + \sin^2\theta\cot^2\theta = 1$
   $\sin^2\theta(1 + \cot^2\theta) = 1$
   $\sin^2\theta(\csc^2\theta) = 1$
   $\sin^2\theta\left(\frac{1}{\sin^2\theta}\right) = 1$
   $1 = 1$

2. $\cot^2\theta\cos^2\theta = \cot^2\theta - \cos^2\theta$

   $\cot^2\theta\cos^2\theta = \cot^2\theta - \cos^2\theta =$
   $\frac{\cos^2\theta}{\sin^2\theta} - \cos^2\theta = \frac{\cos^2\theta}{\sin^2\theta} - \frac{\sin^2\theta\cos^2\theta}{\sin^2\theta} =$
   $\frac{\cos^2\theta - \sin^2\theta\cos^2\theta}{\sin^2\theta} = \frac{\cos^2\theta(1-\sin^2\theta)}{\sin^2\theta} =$
   $\left(\frac{\cos^2\theta}{\sin^2\theta}\right)(1-\sin^2\theta) = \cot^2\theta(1-\sin^2\theta) =$
   $\cot^2\theta(\cos^2\theta + \sin^2\theta - \sin^2\theta) = \cot^2\theta\cos^2\theta$

3. $\tan^2\theta - \tan^2\theta\sin^2\theta = \sin^2\theta$

   $\tan^2\theta - \tan^2\theta\sin^2\theta = \sin^2\theta$
   $\tan^2\theta(1-\sin^2\theta) = \sin^2\theta$
   $\tan^2\theta(\cos^2\theta) = \sin^2\theta$
   $\frac{\sin^2\theta}{\cos^2\theta}(\cos^2\theta) = \sin^2\theta$
   $(\sin^2\theta)\frac{\cos^2\theta}{\cos^2\theta} = \sin^2\theta$
   $\sin^2\theta = \sin^2\theta$

4. $\frac{\sin\theta + \cos\theta}{\sin\theta\cos\theta} = \sec\theta + \csc\theta$

   $\frac{\sin\theta + \cos\theta}{\sin\theta\cos\theta} = \sec\theta + \csc\theta$
   $\frac{\sin\theta}{\sin\theta\cos\theta} + \frac{\cos\theta}{\sin\theta\cos\theta} = \sec\theta + \csc\theta$
   $\left(\frac{\sin\theta}{\sin\theta}\right)\frac{1}{\cos\theta} + \frac{1}{\sin\theta}\left(\frac{\cos\theta}{\cos\theta}\right) = \sec\theta + \csc\theta$
   $(1)\frac{1}{\cos\theta} + \frac{1}{\sin\theta}(1) = \sec\theta + \csc\theta$
   $\frac{1}{\cos\theta} + \frac{1}{\sin\theta} = \sec\theta + \csc\theta$
   $\sec\theta + \csc\theta = \sec\theta + \csc\theta$

Rewrite each expression in terms of cos θ. Then simplify.

5. $2\sin\theta\cos\theta\cot\theta$

   $2\sin\theta\cos\theta\cot\theta =$
   $2\sin\theta\cos\theta \cdot \frac{\cos\theta}{\sin\theta} =$
   $2\cos^2\theta$

6. $\frac{1+\cot\theta}{\cot\theta(\sin\theta + \cos\theta)}$

   $\frac{1+\cot\theta}{\cot\theta(\sin\theta+\cos\theta)} =$
   $\frac{1 + \frac{\cos\theta}{\sin\theta}}{\frac{\cos\theta}{\sin\theta}(\sin\theta+\cos\theta)} =$
   $\frac{\frac{\sin\theta+\cos\theta}{\sin\theta}}{\frac{\cos\theta(\sin\theta+\cos\theta)}{\sin\theta}} =$
   $\frac{\sin\theta+\cos\theta}{\cos\theta(\sin\theta+\cos\theta)} = \frac{1}{\cos\theta}$

7. $\cos^4\theta - \sin^4\theta + \sin^2\theta$

   $\cos^4\theta - \sin^4\theta + \sin^2\theta$
   $= (\cos^2\theta + \sin^2\theta)$
   $(\cos^2\theta - \sin^2\theta) + \sin^2\theta$
   $(1)(\cos^2\theta - \sin^2\theta) +$
   $\sin^2\theta = \cos^2\theta$

Solve.

8. Use the equation $mg\sin\theta = \mu mg\cos\theta$ to determine the angle at which a waxed wood block on an inclined plane of wet snow begins to slide. Assume $\mu = 0.17$. 9.6°

LESSON 14-3

Practice C

Fundamental Trigonometric Identities

Prove each trigonometric identity.

1. $\frac{\cos\theta}{1+\sin\theta} = \frac{1-\sin\theta}{\cos\theta}$

   $\frac{\cos\theta}{1+\sin\theta} = \frac{1-\sin\theta}{\cos\theta}$
   $\frac{\cos\theta}{1+\sin\theta}\left(\frac{1-\sin\theta}{1-\sin\theta}\right) = \frac{1-\sin\theta}{\cos\theta}$
   $\frac{\cos\theta(1-\sin\theta)}{1-\sin^2\theta} = \frac{1-\sin\theta}{\cos\theta}$
   $\frac{\cos\theta(1-\sin\theta)}{\cos^2\theta} = \frac{1-\sin\theta}{\cos\theta}$
   $\frac{1-\sin\theta}{\cos\theta} = \frac{1-\sin\theta}{\cos\theta}$

2. $\tan\theta\sec\theta = \frac{\sec\theta + \tan\theta}{\cos\theta + \cot\theta}$

   $\tan\theta\sec\theta = \frac{\sec\theta + \tan\theta}{\cos\theta + \cot\theta} = \frac{\sec\theta + \frac{\sin\theta}{\cos\theta}}{\cos\theta + \frac{\cos\theta}{\sin\theta}}$
   $= \frac{\frac{1}{\cos\theta} + \frac{\sin\theta}{\cos\theta}}{\cos\theta + \frac{\cos\theta}{\sin\theta}} = \frac{\frac{1}{\cos\theta} + \frac{\sin\theta}{\cos\theta}}{\frac{\sin\theta\cos\theta}{\sin\theta} + \frac{\cos\theta}{\sin\theta}}$
   $= \frac{\frac{1+\sin\theta}{\cos\theta}}{\frac{\sin\theta\cos\theta + \cos\theta}{\sin\theta}} = \frac{\frac{1+\sin\theta}{\cos\theta}}{\frac{\cos\theta(\sin\theta+1)}{\sin\theta}}$
   $= \frac{\sin\theta(1+\sin\theta)}{\cos^2\theta(\sin\theta+1)} = \frac{\sin\theta}{\cos^2\theta}$
   $= \left(\frac{\sin\theta}{\cos\theta}\right)\frac{1}{\cos\theta} = (\tan\theta)\frac{1}{\cos\theta}$
   $= \tan\theta\sec\theta$

3. $\tan^2\theta - \sin^2\theta = \tan^2\theta\sin^2\theta$

   $\tan^2\theta - \sin^2\theta = \tan^2\theta\sin^2\theta$
   $\frac{\sin^2\theta}{\cos^2\theta} - \sin^2\theta = \tan^2\theta\sin^2\theta$
   $\frac{\sin^2\theta - \sin^2\theta\cos^2\theta}{\cos^2\theta} = \tan^2\theta\sin^2\theta$
   $\frac{\sin^2\theta(1-\cos^2\theta)}{\cos^2\theta} = \tan^2\theta\sin^2\theta$
   $\frac{\sin^2\theta(\sin^2\theta)}{\cos^2\theta} = \tan^2\theta\sin^2\theta$
   $\left(\frac{\sin^2\theta}{\cos^2\theta}\right)\sin^2\theta = \tan^2\theta\sin^2\theta$
   $\tan^2\theta\sin^2\theta = \tan^2\theta\sin^2\theta$

4. $1 + \cot^4\theta = \csc^4\theta - 2\cot^2\theta$

   $1 + \cot^4\theta = \csc^4\theta - 2\cot^2\theta$
   $1 + \cot^4\theta = (\cot^2\theta + 1)^2 - 2\cot^2\theta$
   $1 + \cot^4\theta = \cot^4\theta + 2\cot^2\theta + 1 - 2\cot^2\theta$
   $1 + \cot^4\theta = \cot^4\theta + 1$
   $1 + \cot^4\theta = 1 + \cot^4\theta$

Rewrite each expression in terms of sin θ and cos θ. Then simplify.

5. $\frac{\cot\theta}{\sin\theta}$

   $\frac{\cot\theta}{\sin\theta} = \frac{\frac{\cos\theta}{\sin\theta}}{\sin\theta} =$
   $\frac{\cos\theta}{\sin^2\theta} = \frac{\cos\theta}{1-\cos^2\theta}$

6. $\cot^2\theta - \cos\theta\cot^2\theta$

   $\cot^2\theta(1-\cos\theta) =$
   $\frac{\cos^2\theta}{\sin^2\theta}(1-\cos\theta) =$
   $\frac{\cos^2\theta(1-\cos\theta)}{1-\cos^2\theta} =$
   $\frac{\cos^2\theta(1-\cos\theta)}{(1+\cos\theta)(1-\cos\theta)} = \frac{\cos^2\theta}{1+\cos\theta}$

7. $\tan^2\theta + 4\sec^2\theta + 1$

   $\tan^2\theta + 4\sec^2\theta + 1 =$
   $\sec^2\theta - 1 + 4\sec^2\theta + 1 =$
   $5\sec^2\theta = 5\left(\frac{1}{\cos^2\theta}\right) = \frac{5}{\cos^2\theta}$

Solve.

8. Alan is using the equation $mg\sin\theta = \mu mg\cos\theta$ to determine the coefficient of friction, μ, between a flat rock and a metal ramp. Find μ to the nearest hundredth if the rock begins to slide at 19°. 0.34

LESSON 14-3

Reteach

Fundamental Trigonometric Identities

To prove a trigonometric identity, use the fundamental identities to make one side of the equation resemble the other side.

Reciprocal and Ratio Identities		
$\csc\theta = \frac{1}{\sin\theta}$	$\sec\theta = \frac{1}{\cos\theta}$	$\cot\theta = \frac{1}{\tan\theta}$
$\tan\theta = \frac{\sin\theta}{\cos\theta}$		$\cot\theta = \frac{\cos\theta}{\sin\theta}$

Negative-Angle Identities	Pythagorean Identities
$\sin(-\theta) = -\sin\theta$	$\sin^2\theta + \cos^2\theta = 1$
$\cos(-\theta) = \cos\theta$	$\tan^2\theta + 1 = \sec^2\theta$
$\tan(-\theta) = -\tan\theta$	$1 + \cot^2\theta = \csc^2\theta$

As long as you know this identity, you can divide to derive the other two.

Prove: $\tan\theta\sin\theta = \sec\theta - \cos\theta$

Try the right side since you can use reciprocal identities

$\tan\theta\sin\theta = \sec\theta - \cos\theta$	*Modify the right side.*
$= \frac{1}{\cos\theta} - \cos\theta$	*Definition of secant*
$= \frac{1-\cos^2\theta}{\cos\theta}$	*Subtract. Use a common denominator.*
$= \frac{\sin^2\theta}{\cos\theta}$	*Pythagorean identity*
$= \frac{\sin\theta}{\cos\theta}\sin\theta$	*Rewrite the product.*
$= \tan\theta\sin\theta$	*Definition of tangent*

$\sin^2\theta + \cos^2\theta = 1$, so $\sin^2\theta = 1 - \cos^2\theta$.

Write the missing step or reason to prove each trigonometric identity.

1. $\cot(-\theta) = -\cot\theta$ Modify the left side.

   $\frac{\cos(-\theta)}{\sin(-\theta)} = -\cot\theta$ Definition of cotangent

   $\frac{\cos\theta}{\sin(-\theta)} = -\cot\theta$ Negative-angle identity

   $\frac{\cos\theta}{-\sin\theta} = -\cot\theta$ Negative-angle identity

   $-\cot\theta = -\cot\theta$ Definition of cotangent

2. $\sec\theta\cot\theta = \csc\theta$ Modify the left side.

   $\left(\frac{1}{\cos\theta}\right)\left(\frac{\cos\theta}{\sin\theta}\right) = \csc\theta$ Definitions of secant and cotangent

   $\frac{1}{\sin\theta}$ Simplify.

   $\csc\theta = \csc\theta$ Definition of cosecant

LESSON 14-3
Reteach
Fundamental Trigonometric Identities (continued)

Rewriting a trigonometric expression is similar to proving a trigonometric identity. Use the fundamental identities to rewrite the expression.

Trigonometric Identities		
$\csc\theta = \frac{1}{\sin\theta}$	$\sec\theta = \frac{1}{\cos\theta}$	$\cot\theta = \frac{1}{\tan\theta}$
$\tan\theta = \frac{\sin\theta}{\cos\theta}$	$\cot\theta = \frac{\cos\theta}{\sin\theta}$	
$\sin(-\theta) = -\sin\theta$	$\sin^2\theta + \cos^2\theta = 1$	
$\cos(-\theta) = \cos\theta$	$\tan^2\theta + 1 = \sec^2\theta$	
$\tan(-\theta) = -\tan\theta$	$1 + \cot^2\theta = \csc^2\theta$	

Rewrite $\frac{\cos^2\theta}{1-\sin\theta}$ in terms of $\sin\theta$.

$\frac{\cos^2\theta}{1-\sin\theta}$ — $\sin^2\theta + \cos^2\theta = 1$, so $\cos^2\theta = 1 - \sin^2\theta$.

$\frac{1-\sin^2\theta}{1-\sin\theta}$ — *Pythagorean identity*

$\frac{(1-\sin\theta)(1+\sin\theta)}{1-\sin\theta}$ — *Factor the numerator.* Think: Let $a = \sin\theta$. $1 - a^2 = (1-a)(1+a)$

$\frac{\cancel{(1-\sin\theta)}(1+\sin\theta)}{\cancel{1-\sin\theta}}$ — *Simplify.*

$1 + \sin\theta$ — The expression is in terms of $\sin\theta$.

Write the missing step or reason to rewrite each expression in terms of $\cos\theta$.

3. $(1 + \sec^2\theta)(1 - \sin^2\theta)$

$\left(1 + \frac{1}{\cos^2\theta}\right)(1 - \sin^2\theta)$ — Definition of secant

$\left(1 + \frac{1}{\cos^2\theta}\right)(\cos^2\theta)$ — Pythagorean identity

$\cos^2\theta + 1$ — Multiply.

4. $\tan^2\theta$

$\frac{\sin^2\theta}{\cos^2\theta}$ — Definition of tangent

$\frac{1-\cos^2\theta}{\cos^2\theta}$ — Pythagorean identity

$\frac{1}{\cos^2\theta} - 1$ — Use the Distributive Property.

LESSON 14-3
Challenge
Expressing One Function in Terms of Another

Equations that are true for all values of the variables for which they are defined are called *identities*. There are three groups of basic trigonometric identities.

Reciprocal Identities	Ratio Identities	Pythagorean Identities
$\csc\theta = \frac{1}{\sin\theta}$ with $\sin\theta \neq 0$	$\tan\theta = \frac{\sin\theta}{\cos\theta}$ with $\cos\theta \neq 0$	$\sin^2\theta + \cos^2\theta = 1$
$\sec\theta = \frac{1}{\cos\theta}$ with $\cos\theta \neq 0$	$\cot\theta = \frac{\cos\theta}{\sin\theta}$ with $\sin\theta \neq 0$	$\tan^2\theta + 1 = \sec^2\theta$
$\cot\theta = \frac{1}{\tan\theta}$ with $\tan\theta \neq 0$		$\cot^2\theta + 1 = \csc^2\theta$

You can use these identities to express one trigonometric function in terms of another.

To express $\tan\theta$ in terms of $\sin\theta$, begin with the ratio identity for $\tan\theta$. $\tan\theta = \frac{\sin\theta}{\cos\theta}$

Now find $\cos\theta$ in terms of $\sin\theta$. Use the appropriate Pythagorean identity, and solve for $\cos\theta$. $\sin^2\theta + \cos^2\theta = 1$; $\cos^2\theta = 1 - \sin^2\theta$; $\cos\theta = \pm\sqrt{1-\sin^2\theta}$

Return to the ratio identity and substitute for $\cos\theta$. $\tan\theta = \frac{\sin\theta}{\pm\sqrt{1-\sin^2\theta}}$

Express the given function in terms of $\cos\theta$.

1. $\tan\theta$ — $\frac{\pm\sqrt{1-\cos^2\theta}}{\cos\theta}$
2. $\cot\theta$ — $\frac{\cos\theta}{\pm\sqrt{1-\cos^2\theta}}$
3. $\csc\theta$ — $\frac{1}{\pm\sqrt{1-\cos^2\theta}}$

Express the given function in terms of $\tan\theta$.

4. $\cot\theta$ — $\frac{1}{\tan\theta}$
5. $\sin\theta$ — $\frac{\tan\theta}{\pm\sqrt{1+\tan^2\theta}}$
6. $\cos\theta$ — $\frac{1}{\pm\sqrt{1+\tan^2\theta}}$

Express the given function in terms of $\csc\theta$.

7. $\cos\theta$ — $\frac{\pm\sqrt{\csc^2\theta-1}}{\csc\theta}$
8. $\tan\theta$ — $\frac{1}{\pm\sqrt{\csc^2\theta-1}}$
9. $\sec\theta$ — $\frac{\csc\theta}{\pm\sqrt{\csc^2\theta-1}}$

Express the given function in terms of $\sec\theta$.

10. $\tan\theta$ — $\pm\sqrt{\sec^2\theta-1}$
11. $\sin\theta$ — $\frac{\pm\sqrt{\sec^2\theta-1}}{\sec\theta}$
12. $\cot\theta$ — $\frac{1}{\pm\sqrt{\sec^2\theta-1}}$

LESSON 14-3
Problem Solving
Fundamental Trigonometric Identities

The advertisement for a new shoe promises runners less slip. The coefficient of friction (μ) between concrete and a new material used in the sole of this shoe is 1.5. The force of friction that causes slip is equal to $mg\sin\theta$, where m is a runner's mass and g is the acceleration due to gravity. The force that prevents slip is $\mu mg\cos\theta$.

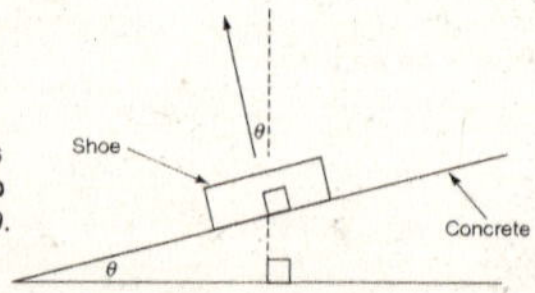

1. Mukisa wants to know what this new shoe will do for his performance.
   a. At the instant of slip, the force that causes slip is equal to the force that prevents it. Write an equation to show this relationship. — $mg\sin\theta = \mu mg\cos\theta$
   b. Rewrite the equation with trigonometric expressions on one side of the equation, and substitute the coefficient of friction for the new material. — $\frac{\sin\theta}{\cos\theta} = 1.5$
   c. Use a trigonometric identity to rewrite the equation using only the tangent function. — $\tan\theta = 1.5$
   d. Solve for θ, the angle at which the new shoe will start to slip. — 56.3°
2. Mukisa wonders how his old shoes compare to these new shoes. The label on the box of his old shoes states that the coefficient of friction is 1.3.
   a. Write an equation to find the angle at which a shoe with a coefficient of friction of 1.3 will slip. — $mg\sin\theta = (1.3)mg\cos\theta$
   b. At what angle will this shoe start to slip? — 52.4°
   c. Suggest how the new shoes might improve Mukisa's performance.

   Possible answer: With the new shoes he should be able to push off at a greater angle (by about 4°) to the vertical.

An advertisement states, "Use this wax and your skis will slide better than silk on silk!" The coefficient of friction for silk on silk is 0.25. The coefficient of friction for waxed wood on wet snow is 0.14. The equation $mg\sin\theta = \mu mg\cos\theta$ can be used to find the angle at which a material begins to slide. Choose the letter for the best answer.

3. Which expression gives the angle at which silk begins to slide on silk?
   A $\cos^{-1}(0.25)$
   B $\sin\left(\frac{1}{0.25}\right)$
   C $\tan\left(\frac{1}{0.25}\right)$
   (D) $\tan^{-1}(0.25)$

4. At what angle would a waxed wooden ski begin to slide on wet snow?
   F 2°
   (G) 8°
   H 12°
   J 14°

LESSON 14-3
Reading Strategy
Identify Relationships

The trigonometric identities are relationships that we can use to rewrite and simplify trigonometric functions.

Reciprocal Identities	Ratio Identities
$\csc\theta = \frac{1}{\sin\theta}$	$\tan\theta = \frac{\sin\theta}{\cos\theta}$
$\sec\theta = \frac{1}{\cos\theta}$	$\cot\theta = \frac{\cos\theta}{\sin\theta}$
$\cot\theta = \frac{1}{\tan\theta}$	

Answer each question.

1. Which identities are like inverse relationships? Explain.

   Possible answer: the cosecant, secant, and cotangent functions because they are the opposite of the sine, cosine, and tangent functions

2. Rewrite each as its reciprocal identity.
   a. sec85° — $\frac{1}{\cos 85°}$
   b. csc150° — $\frac{1}{\sin 150°}$
   c. cot22° — $\frac{1}{\tan 22°}$

3. Rewrite each as its ratio identity.
   a. tan340° — $\frac{\sin 340°}{\cos 340°}$
   b. $\frac{1}{\tan 230°}$ — $\frac{\cos 230°}{\sin 230°}$
   c. $\frac{1}{\cot 45°}$ — $\frac{\sin 45°}{\cos 45°}$

Rewrite each expression (a) using reciprocal identities, and then (b) simplify using ratio identities.

4. $\csc\theta - \cot\theta$ — a. $\frac{1}{\sin\theta} - \frac{1}{\tan\theta}$ — b. $\frac{1-\cos\theta}{\sin\theta}$
5. $\sec\theta + \frac{1}{\cot\theta}$ — a. $\frac{1}{\cos\theta} + \tan\theta$ — b. $\frac{1+\sin\theta}{\cos\theta}$
6. $\cot\theta + 2(\csc\theta)$ — a. $\frac{1}{\tan\theta} + \frac{2}{\sin\theta}$ — b. $\frac{\cos\theta + 2}{\sin\theta}$

Rewrite each expression in terms of sine and cosine using reciprocal identities and ratio identities.

7. $\frac{1}{\cot\theta}$ — $\frac{\sin\theta}{\cos\theta}$
8. $\frac{1}{\csc\theta} + \tan\theta$ — $\sin\theta + \frac{\sin\theta}{\cos\theta}$
9. $\frac{1}{\tan\theta} + \csc\theta$ — $\frac{\cos\theta + 1}{\sin\theta}$
10. $\frac{1}{\sec\theta} + \tan\theta$ — $\cos\theta + \frac{\sin\theta}{\cos\theta}$

LESSON 14-4

Practice A

Sum and Difference Identities

Find the exact value of each expression.

1. cos210°

   a. Write 210° as a sum or difference of two known trigonometric values. $180° + 30°$

   b. Substitute the values into the correct sum or difference identity.
   $\cos(180° + 30°) = \cos180°\cos30° - \sin180°\sin30°$

   c. Evaluate and simplify. $-1 \cdot \frac{\sqrt{3}}{2} - 0 \cdot \frac{1}{2} = -\frac{\sqrt{3}}{2}$

2. cos315° $\frac{\sqrt{2}}{2}$

3. sin105° $\frac{\sqrt{6} + \sqrt{2}}{4}$

4. tan15° $2 - \sqrt{3}$

5. $\tan\frac{5\pi}{12}$ $2 + \sqrt{3}$

6. $\sin\frac{5\pi}{3}$ $-\frac{\sqrt{3}}{2}$

7. $\cos\frac{2\pi}{3}$ $-\frac{1}{2}$

Let $\sin A = \frac{12}{13}$ with $90° \le A \le 180°$ and $\tan B = -\frac{4}{3}$ with $270° \le B \le 360°$. Find each value.

8. $\tan(A + B)$

   a. Use $\sin A = \frac{12}{13}$ to find $\tan A$. $\tan A = -\frac{12}{5}$

   b. Substitute known values into the angle sum identity $\tan(A + B) = \frac{\tan A + \tan B}{1 - \tan A \tan B}$.
   $\tan(A + B) = \frac{-\frac{12}{5} + \left(-\frac{4}{3}\right)}{1 - \left(-\frac{12}{5}\left(-\frac{4}{3}\right)\right)}$

   c. Simplify. $\tan(A + B) = \frac{56}{33}$

9. $\tan(A - B)$ $-\frac{16}{63}$

10. $\sin(A + B)$ $\frac{56}{65}$

11. $\cos(A - B)$ $-\frac{63}{65}$

Solve.

12. The displacement, y, of a mass attached to a spring is modeled by $y(t) = 9.6\cos\left(\frac{2\pi}{6}(t) - \frac{\pi}{2}\right)$, where t is the time in seconds.

    a. What are the amplitude and period of the function? Amplitude: 9.6; period: 6

    b. Write the displacement using only the sine function. $y(t) = 9.6\sin\left(\frac{2\pi}{6}(t)\right)$

    c. What is the displacement of mass when $t = 6$? 0

LESSON 14-4

Practice B

Sum and Difference Identities

Find the exact value of each expression.

1. cos120° $-\frac{1}{2}$

2. sin315° $-\frac{\sqrt{2}}{2}$

3. tan255° $2 + \sqrt{3}$

4. $\tan\frac{7\pi}{6}$ $\frac{\sqrt{3}}{3}$

5. $\sin\frac{\pi}{12}$ $\frac{\sqrt{6} - \sqrt{2}}{4}$

6. $\cos\frac{3\pi}{4}$ $-\frac{\sqrt{2}}{2}$

Prove each identity.

7. $\sin\left(x - \frac{3\pi}{2}\right) = \cos x$

   $\sin\left(x - \frac{3\pi}{2}\right) = \cos x;$
   $\sin x\cos\left(\frac{3\pi}{2}\right) - \sin\left(\frac{3\pi}{2}\right)\cos x = \cos x$
   $\sin x \cdot 0 - (-1)\cos x = \cos x;$
   $\cos x = \cos x$

8. $\cos\left(x - \frac{\pi}{2}\right) = \sin x$

   $\cos\left(x - \frac{\pi}{2}\right) = \sin x$
   $\cos x\cos\left(\frac{\pi}{2}\right) + \sin x\sin\left(\frac{\pi}{2}\right) = \sin x$
   $\cos x \cdot 0 + \sin x \cdot 1 = \sin x$
   $\sin x = \sin x$

Find each value if $\cos A = \frac{12}{13}$ with $0° \le A \le 90°$ and if $\sin B = \frac{8}{17}$ with $90° \le B \le 180°$.

9. $\sin(A + B)$ $\frac{21}{221}$

10. $\cos(A + B)$ $-\frac{220}{221}$

11. $\tan(A + B)$ $-\frac{21}{220}$

12. $\sin(A - B)$ $-\frac{171}{221}$

13. $\cos(A - B)$ $-\frac{140}{221}$

14. $\tan(A - B)$ $\frac{171}{140}$

Solve.

15. Find the coordinates, to the nearest hundredth, of the vertices of triangle ABC with $A(1, 0)$, $B(10, 0)$, and $C(2, 6)$ after a 60° rotation about the origin.

    a. Write the matrices for the rotation and for the points. $\begin{bmatrix} \cos 60° & -\sin 60° \\ \sin 60° & \cos 60° \end{bmatrix}\begin{bmatrix} 1 & 10 & 2 \\ 0 & 0 & 6 \end{bmatrix}$

    b. Find the matrix product. $\begin{bmatrix} 0.50 & 5.00 & -4.20 \\ 0.87 & 8.66 & 4.73 \end{bmatrix}$

    c. Write the coordinates. $A'(0.50, 0.87)$, $B'(5, 8.66)$, $C'(-4.2, 4.73)$

16. A hill rises from the horizontal at a 15° angle. The road leading straight up the hill is 800 meters long. How much higher is the top of the hill than the base of the hill? 207 m

LESSON 14-4

Practice C

Sum and Difference Identities

Find the exact value of each expression.

1. cos300° $\frac{1}{2}$

2. sin(−15°) $\frac{\sqrt{2} - \sqrt{6}}{4}$

3. tan285° $-2 - \sqrt{3}$

4. $\tan\frac{11\pi}{12}$ $-2 + \sqrt{3}$

5. $\sin\frac{13\pi}{12}$ $\frac{\sqrt{2} - \sqrt{6}}{4}$

6. $\cos\left(-\frac{13\pi}{12}\right)$ $-\frac{\sqrt{6} - \sqrt{2}}{4}$

Prove each identity.

7. $\tan(\pi - x) = -\tan x$

   $\tan(\pi - x) = -\tan x$
   $\frac{\tan\pi - \tan x}{1 + \tan\pi\tan x} = -\tan x$
   $\frac{0 - \tan x}{1 + 0 \cdot \tan x} = -\tan x$
   $-\tan x = -\tan x$

8. $\sin\left(\frac{\pi}{2} - x\right) = \cos x$

   $\sin\left(\frac{\pi}{2} - x\right) = \cos x$
   $\sin\left(\frac{\pi}{2}\right)\cos x - \cos\left(\frac{\pi}{2}\right)\sin x = \cos x$
   $1 \cdot \cos x - 0 \cdot \sin x = \cos x$
   $\cos x = \cos x$

Find each value if $\sin A = \frac{12}{13}$ with $90° \le A \le 180°$ and if $\sin B = -\frac{7}{25}$ with $-90° \le B \le 0°$.

9. $\sin(A + B)$ $\frac{323}{325}$

10. $\cos(A + B)$ $-\frac{36}{325}$

11. $\tan(A + B)$ $-\frac{323}{36}$

12. $\sin(A - B)$ $\frac{253}{325}$

13. $\cos(A - B)$ $-\frac{204}{325}$

14. $\tan(A - B)$ $-\frac{253}{204}$

Solve.

15. Find the coordinates, to the nearest hundredth, of the vertices of triangle PQR with $P(0, 1)$, $Q(0, -4)$, and $R(2, 5)$ after a 45° rotation about the origin.
    $P'(-0.71, 0.71)$, $Q'(2.83, -2.83)$, $R'(-2.12, 4.95)$

16. Find the coordinates, to the nearest hundredth, of the vertices of triangle FGH with $F(0, 10)$, $G(10, 0)$, and $H(1, -1)$ after a 75° rotation about the origin.
    $F'(-9.66, 2.59)$, $G'(2.59, 9.66)$, $H'(1.22, 0.71)$

17. Two pilots take off from the same airport. Mason heads due south. Nancy heads 23° west of south. After 400 land miles, how far is Nancy from Mason's route? 156.3 mi

LESSON 14-4

Reteach

Sum and Difference Identities

You can use angle addition and subtraction identities to find the exact value of some trigonometric expressions.

Look for ways to use 30° or $\frac{\pi}{6}$, 45° or $\frac{\pi}{4}$, and 60° or $\frac{\pi}{3}$ to make the sum because exact values are known for these angles.

Sum Identities	Difference Identities
$\sin(A + B) = \sin A\cos B + \cos A\sin B$	$\sin(A - B) = \sin A\cos B - \cos A\sin B$
$\cos(A + B) = \cos A\cos B - \sin A\sin B$	$\cos(A - B) = \cos A\cos B + \sin A\sin B$
$\tan(A + B) = \frac{\tan A + \tan B}{1 - \tan A\tan B}$	$\tan(A - B) = \frac{\tan A - \tan B}{1 + \tan A\tan B}$

Find the exact value of cos105°.

$\cos105° = \cos(60° + 45°)$ — *Think: 60° + 45° = 105°. Use cos(A + B) identity.*
$= \cos60°\cos45° - \sin60°\sin45°$
$= \frac{1}{2} \cdot \frac{\sqrt{2}}{2} - \frac{\sqrt{3}}{2} \cdot \frac{\sqrt{2}}{2}$ — *Evaluate.*
$= \frac{\sqrt{2}}{4} - \frac{\sqrt{6}}{4} = \frac{\sqrt{2} - \sqrt{6}}{4}$ — *Simplify.*

Substitute: $A = 60°$ and $B = 45°$.

The value is negative. This makes sense since 105° lies in Quadrant II where cosine is negative.

Find the exact value of $\sin\left(\frac{\pi}{12}\right)$.

$\sin\left(\frac{\pi}{12}\right) = \sin\left(\frac{\pi}{3} - \frac{\pi}{4}\right)$ — *Think: $\frac{\pi}{3} - \frac{\pi}{4} = \frac{\pi}{12}$.*
$= \sin\frac{\pi}{3}\cos\frac{\pi}{4} - \cos\frac{\pi}{3}\sin\frac{\pi}{4}$ — *Use sin(A − B) identity.*
$= \frac{\sqrt{3}}{2} \cdot \frac{\sqrt{2}}{2} - \frac{1}{2} \cdot \frac{\sqrt{2}}{2}$ — *Evaluate.*
$= \frac{\sqrt{6}}{4} - \frac{\sqrt{2}}{4} = \frac{\sqrt{6} - \sqrt{2}}{4}$ — *Simplify.*

Substitute: $A = \frac{\pi}{3}$ and $B = \frac{\pi}{4}$.

Use the sum or difference identity to find the exact value of each expression.

1. $\cos(-15°) = \cos(30° - 45°)$

   $\cos30°\cos45° + \sin30°\sin45°$
   $= \frac{\sqrt{3}}{2} \cdot \frac{\sqrt{2}}{2} + \frac{1}{2} \cdot \frac{\sqrt{2}}{2}$
   $= \frac{\sqrt{6} + \sqrt{2}}{4}$

2. $\sin\left(\frac{7\pi}{12}\right) = \sin\left(\frac{\pi}{3} + \frac{\pi}{4}\right)$

   $\sin\frac{\pi}{3}\cos\frac{\pi}{4} + \cos\frac{\pi}{3}\sin\frac{\pi}{4}$
   $= \frac{\sqrt{3}}{2} \cdot \frac{\sqrt{2}}{2} + \frac{1}{2} \cdot \frac{\sqrt{2}}{2}$
   $= \frac{\sqrt{6} + \sqrt{2}}{4}$

LESSON 14-4
Reteach
Sum and Difference Identities (continued)

You can use angle addition and subtraction identities to prove identities. Use an identity to make one side of the equation resemble the other side.

Sum Identities	Difference Identities
$\sin(A+B) = \sin A\cos B + \cos A\sin B$	$\sin(A-B) = \sin A\cos B - \cos A\sin B$
$\cos(A+B) = \cos A\cos B - \sin A\sin B$	$\cos(A-B) = \cos A\cos B + \sin A\sin B$
$\tan(A+B) = \dfrac{\tan A + \tan B}{1 - \tan A\tan B}$	$\tan(A-B) = \dfrac{\tan A - \tan B}{1 + \tan A\tan B}$

Prove: $\tan(\pi + x) = \tan x$

$\tan(\pi + x) = \tan x$	*Modify the left side.*
$\dfrac{\tan\pi + \tan x}{1 - \tan\pi\tan x} = \tan x$	*Use tan(A + B) identity.* (Substitute: $A = \pi$ and $B = x$.)
$\dfrac{0 + \tan x}{1 - (0)\tan x} = \tan x$	*Evaluate. Think: $\tan\pi = 0$.*
$\dfrac{\tan x}{1} = \tan x$	*Simplify.*

Prove: $\sin(\pi - x) = \sin x$

$\sin(\pi - x) = \sin x$	*Modify the left side.*
$\sin\pi\cos x - \cos\pi\sin x = \sin x$	*Use sin(A − B) identity.*
$(0)\cos x - (-1)\sin x = \sin x$	*Evaluate. Think: $\sin\pi = 0$ and $\cos\pi = -1$.*
$0 + \sin x = \sin x$	*Simplify.*

Write the missing steps or reasons to prove each identity.

3. $\sin\left(\frac{\pi}{2} + x\right) = \cos x$ — Modify the left side.

$\sin\left(\frac{\pi}{2}\right)\cos x + \cos\left(\frac{\pi}{2}\right)\sin x = \cos x$ — **Use $\sin(A + B)$ identity**

$(1)(\cos x) + 0(\sin x) = \cos x$ — Evaluate.

$\cos x = \cos x$ — Simplify.

4. $\cos(\pi + x) = -\cos x$ — Modify the left side.

$\cos\pi\cos x - \sin\pi\sin x = -\cos x$ — Use cos (A + B) identity.

$(-1)\cos x - (0)(\sin x) = -\cos x$ — **Evaluate.**

$-\cos x = -\cos x$ — **Simplify.**

LESSON 14-4
Challenge
Proof of Identity

The diagram at right provides a geometric proof of the trigonometric identity $\sin(x + y) = \sin x\cos y + \cos x\sin y$. Angles x and y are located at the origin with angle y in standard position and the initial side of angle x located on the terminal side of angle y. The circle has $r = 1$. All other sides of the triangle are marked.

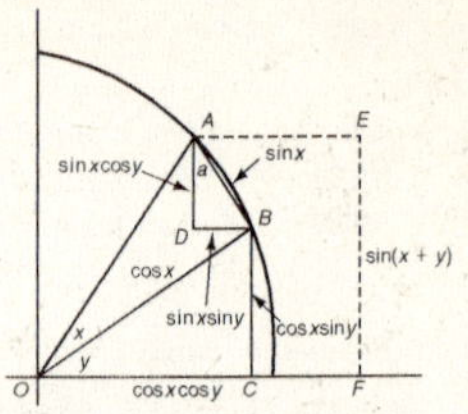

1. Using the properties of right triangles and the definitions of the trigonometric functions, justify the lengths of the sides of triangles *AOB* and *BOC*.

Possible answer: Since the triangles are right triangles and the circle has a radius of 1, the length of the side opposite angle x is $\sin x$ and the length of the adjacent side is $\cos x$. In the right triangle with angle y, the length of the hypotenuse is $\cos x$ so the length of the side opposite angle y is $\cos x\sin y$ and the adjacent side is $\cos x\cos y$.

2. Prove that angle a in triangle *ABD* is equal in measure to angle y.

Possible answer: $m\angle OAD + a = 90° - x$; $x + y = 90° - m\angle OAD$; so $a = 90° - x - m\angle OAD$ and $y = 90° - m\angle OAD - x$; therefore, $a = y$.

3. Using the properties of right triangles and the definitions of the trigonometric functions, justify the lengths of the sides of triangle *ABD*.

Possible answer: Triangle ABD is a right triangle, therefore, $AD = AB\cos a$ and $BD = AB\sin a$. But $AB = \sin x$ and $a = y$, so $AD = \sin x\cos y$ and $BD = \sin x\sin y$.

4. Now prove the identity for the sine of a sum, $\sin(x + y) = \sin x\cos y + \cos x\sin y$.

Possible answer: In right triangle OBC, $OB = \cos x$ and $BC = OB\sin y = \cos x\sin y$. From the diagram, it is clear why the length EF is, as the label indicates, equal to $\sin(x + y)$. But $EF = AD + BC$. Substitution of the expressions for EF, AD, and BC gives $\sin(x + y) = \sin x\cos y + \cos x\sin y$.

5. Use this information to prove the sum identity for cosines, $\cos(x + y) = \cos x\cos y - \sin x\sin y$.

Possible answer: Extend segment AD to intersect segment OC at a point P. The segment AP will be perpendicular to segment OC so it follows that $OP = \cos(x + y)$. But $OP + PC = OC$, and hence $OP = OC - PC$. Since $OC = \cos x\cos y$ and $PC = BD = \sin x\sin y$, it follows that $\cos(x + y) = \cos x\cos y - \sin x\sin y$.

LESSON 14-4
Problem Solving
Sum and Difference Identities

Caitlin is designing a rotating graphic for her website. She explores what it will look like if it flashes in different positions.

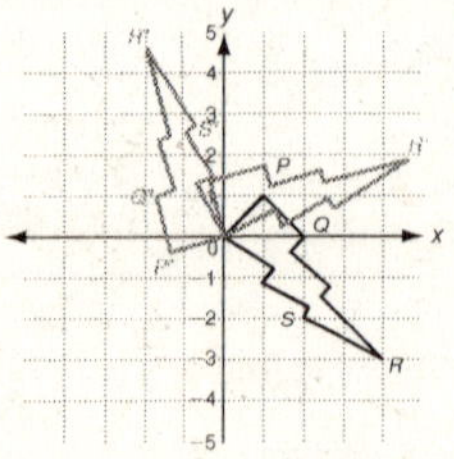

1. Use rotation matrices to find the coordinates of the figure *PQRS* after a 60° rotation about the origin.

a. Write matrices for a 60° rotation and for the points *P*, *Q*, *R*, and *S* in the figure.

$$R_{60°} = \begin{bmatrix} \cos 60° & -\sin 60° \\ \sin 60° & \cos 60° \end{bmatrix}$$

$$PQRS = \begin{bmatrix} 1 & 2 & 4 & 2 \\ 1 & 0 & -3 & -2 \end{bmatrix}$$

b. Find the matrix product. $R_{60°} \times PQRS \approx \begin{bmatrix} -0.37 & 1 & 4.60 & 2.73 \\ 1.37 & 1.73 & 1.96 & 0.73 \end{bmatrix}$

c. Name the coordinates of P', Q', R', S' to the nearest hundredth after a 60° rotation.

$P'(-0.37, 1.37)$, $Q'(1, 1.73)$, $R'(4.60, 1.96)$, and $S'(2.73, 0.73)$

2. a. Write the matrix product that gives the coordinates of the figure after a rotation of 150°.

$$R_{150°} \times PQRS = \begin{bmatrix} -1.37 & -1.73 & -1.96 & -0.73 \\ -0.37 & 1 & 4.60 & 2.73 \end{bmatrix}$$

b. Name the coordinates of $P''Q''R''S''$ to the nearest hundredth.

$P''(-1.37, -0.37)$, $Q''(-1.73, 1)$, $R''(-1.96, 4.60)$, and $S''(-0.73, 2.73)$

3. Graph the two rotations on the plane above to see where Caitlin's rotating graphic will flash. Label R' and R''.

Choose the letter for the best answer.

4. Which expression gives the exact value of sin150°?
   - **(A)** sin90°cos60° + cos90°sin60°
   - B sin90°cos60° − cos90°sin60°
   - C cos180°cos30° + sin180°sin30°
   - D cos180°cos30° − sin180°sin30°

5. What is the exact value of cos75°?
   - F $\dfrac{\sqrt{2} - \sqrt{6}}{2}$
   - G $\dfrac{\sqrt{6} + \sqrt{2}}{4}$
   - **(H)** $\dfrac{\sqrt{6} - \sqrt{2}}{4}$
   - J $\dfrac{\sqrt{6} + \sqrt{2}}{2}$

LESSON 14-4
Reading Strategy
Follow a Procedure

You can use sum and difference identities, along with the Unit Circle, to simplify expressions and find solutions.

Sine Sum Identity	Sine Difference Identity
$\sin(A + B) = \sin A\cos B + \cos A\sin B$	$\sin(A - B) = \sin A\cos B - \cos A\sin B$

To find the value of sin105°:

STEP 1: Determine whether the angle is the sum of two known angles, or the difference: 105° = 45° + 60°

STEP 2: Substitute the new angles into the sum or difference identity: sin(105°) = sin45°cos60° + cos45°sin60°

STEP 3: Solve:
sin(105°) = sin45°cos60° + cos45°sin60°

$$= \frac{\sqrt{2}}{2} \cdot \frac{1}{2} + \frac{\sqrt{2}}{2} \cdot \frac{\sqrt{3}}{2}$$

$$= \frac{\sqrt{2}}{4} + \frac{\sqrt{6}}{4} = \frac{\sqrt{2} + \sqrt{6}}{4}$$

Identify a pair of angles whose trigonometric values are known and whose sum or difference is equal to each angle given.

1. 75° **30° + 45°**
2. 255° **210° + 45°**
3. 165° **135° + 30°**
4. 105° **150° − 45°**

Write the following as sum or difference identities.

5. sin(45° − 30°) **sin45°cos30° − cos45°sin30°**
6. sin(240° + 60°) **sin240°cos60° + cos240°sin60°**
7. sin(300° − 210°) **sin300°cos210° − cos300°sin210°**
8. sin(180° + 45°) **sin180°cos45° + cos180°sin45°**
9. sin(90° − 60°) **sin90°cos60° − cos90°sin60°**
10. sin(225° − 135°) **sin225°cos135° − cos225°sin135°**

LESSON 14-5
Practice A
Double-Angle and Half-Angle Identities

Find sin 2θ, cos 2θ, and tan 2θ for each set of conditions.

1. $\sin\theta = -\frac{24}{25}$ for $\pi < \theta < \frac{3\pi}{2}$
   a. Find cos θ by substituting into formulas for sin 2θ and cos 2θ. $-\frac{7}{25}$
   b. Use the double-angle identity $\sin 2\theta = 2\sin\theta\cos\theta$ to evaluate sin 2θ. $\frac{336}{625}$
   c. Use the double-angle identity $\cos 2\theta = \cos^2\theta - \sin^2\theta$ to evaluate cos 2θ. $-\frac{527}{625}$
   d. Find tan θ for substituting into formula for tan 2θ. $\frac{24}{7}$
   e. Use $\tan 2\theta = \frac{2\tan\theta}{1-\tan^2\theta}$ to evaluate tan 2θ. $-\frac{336}{527}$

2. $\cos\theta = -\frac{3}{5}$ for $\frac{\pi}{2} < \theta < \pi$
   $-\frac{24}{25}, -\frac{7}{25}, \frac{24}{7}$

3. $\sin\theta = \frac{5}{8}$ for $0 < \theta < \frac{\pi}{2}$
   $\frac{5\sqrt{39}}{32}, \frac{7}{32}, \frac{5\sqrt{39}}{7}$

4. $\sin\theta = -\frac{1}{6}$ for $\frac{3\pi}{2} < \theta < 2\pi$
   $-\frac{\sqrt{35}}{18}, \frac{17}{18}, -\frac{\sqrt{35}}{17}$

5. $\tan\theta = \frac{1}{3}$ for $\pi < \theta < \frac{3\pi}{2}$
   $\frac{3}{5}, \frac{4}{5}, \frac{3}{4}$

Prove each identity.

6. $\sin 2\theta + 1 = (\sin\theta + \cos\theta)^2$
   a. Expand the right side of the equation. $\sin^2\theta + 2\sin\theta\cos\theta + \cos^2\theta$
   b. Use the Associative Property of Addition. $2\sin\theta\cos\theta + \sin^2\theta + \cos^2\theta$
   c. Use the double-angle identity for sin 2θ. $\sin 2\theta + \sin^2\theta + \cos^2\theta$
   d. Use the Pythagorean Identity. $\sin 2\theta + 1$

7. $\frac{\sin 2\theta}{2\cos^2\theta} = \tan\theta$

   $\frac{\sin 2\theta}{2\cos^2\theta} = \tan\theta$; $\frac{2\sin\theta\cos\theta}{2\cos^2\theta} = \tan\theta$;
   $\frac{\sin\theta}{\cos\theta} = \tan\theta$; $\tan\theta = \tan\theta$

8. $2\cos^2\theta - 1 = \cos 2\theta$

   $2\cos^2\theta - 1 = \cos 2\theta$
   $2\cos^2\theta - 1 = \cos^2\theta - \sin^2\theta$
   $2\cos^2\theta - 1 = \cos^2\theta - (1 - \cos^2\theta)$
   $2\cos^2\theta - 1 = \cos^2\theta - 1 + \cos^2\theta$
   $2\cos^2\theta - 1 = 2\cos^2\theta - 1$

LESSON 14-5
Practice B
Double-Angle and Half-Angle Identities

Find sin 2θ, cos 2θ, and tan 2θ for each.

1. $\cos\theta = -\frac{12}{13}$ for $\pi < \theta < \frac{3\pi}{2}$
   $\frac{120}{169}, \frac{119}{169}, \frac{120}{119}$

2. $\sin\theta = \frac{\sqrt{6}}{10}$ for $0 < \theta < \frac{\pi}{2}$
   $\frac{\sqrt{141}}{25}, \frac{22}{25}, \frac{\sqrt{141}}{22}$

3. $\sin\theta = -\frac{2}{3}$ for $\frac{3\pi}{2} < \theta < 2\pi$
   $-\frac{4\sqrt{5}}{9}, \frac{1}{9}, -4\sqrt{5}$

4. $\tan\theta = -\frac{5}{6}$ for $\frac{\pi}{2} < \theta < \pi$
   $-\frac{60}{61}, \frac{11}{61}, -\frac{60}{11}$

Prove each identity.

5. $2\cos^2\theta = \cos 2\theta + 1$

   $2\cos^2\theta = \cos 2\theta + 1$
   $2\cos^2\theta = \cos^2\theta - \sin^2\theta + 1$
   $2\cos^2\theta = \cos^2\theta - (1 - \cos^2\theta) + 1$
   $2\cos^2\theta = \cos^2\theta - 1 + \cos^2\theta + 1$
   $2\cos^2\theta = 2\cos^2\theta$

6. $\tan\theta = \frac{1 - \cos 2\theta}{\sin 2\theta}$

   $\tan\theta = \frac{1 - \cos 2\theta}{\sin 2\theta}$;
   $\tan\theta = \frac{1 - (1 - 2\sin^2\theta)}{2\sin\theta\cos\theta}$;
   $\tan\theta = \frac{1 - 1 + 2\sin^2\theta}{2\sin\theta\cos\theta}$;
   $\tan\theta = \frac{\sin\theta}{\cos\theta}$; $\tan\theta = \tan\theta$

Use half-angle identities to find the exact value of each trigonometric expression.

7. tan 22.5°
   $\sqrt{\frac{2-\sqrt{2}}{2+\sqrt{2}}}$

8. $\cos\frac{7\pi}{12}$
   $-\frac{\sqrt{2-\sqrt{3}}}{2}$

9. $\sin\frac{11\pi}{12}$
   $\frac{\sqrt{2-\sqrt{3}}}{2}$

Find $\sin\frac{\theta}{2}$, $\cos\frac{\theta}{2}$, and $\tan\frac{\theta}{2}$ for each.

10. $\cos\theta = \frac{3}{5}$ and $270° < \theta < 360°$
    $\frac{\sqrt{5}}{5}, -\frac{2\sqrt{5}}{5}, -\frac{1}{2}$

11. $\sin\theta = -\frac{\sqrt{5}}{3}$ and $180° < \theta < 270°$
    $\frac{\sqrt{30}}{6}, -\frac{\sqrt{6}}{6}, -\sqrt{5}$

Solve.

12. A water-park slide covers 100 feet of horizontal space and is 36 feet high.
    a. Write a trigonometric relation in terms of θ, the angle that the slide makes with the water surface. $\tan\theta = \frac{9}{25}$
    b. A new replacement slide will create an angle with the water surface that measures twice that of the original slide. The new slide will use the same horizontal space as the old slide. Write an expression that can be evaluated to find the height of the new slide. $\frac{100 \cdot 2\left(\frac{9}{25}\right)}{1 - \left(\frac{9}{25}\right)^2}$
    c. What is the height of the new slide to the nearest foot? 83 ft

LESSON 14-5
Practice C
Double-Angle and Half-Angle Identities

Find sin 2θ, cos 2θ, and tan 2θ for each.

1. $\cos\theta = \frac{\sqrt{3}}{10}$ for $\frac{3\pi}{2} < \theta < 2\pi$
   $-\frac{\sqrt{291}}{50}, -\frac{47}{50}, \frac{\sqrt{291}}{47}$

2. $\tan\theta = \frac{4}{7}$ for $\pi < \theta < \frac{3\pi}{2}$
   $\frac{56}{65}, \frac{33}{65}, \frac{56}{33}$

3. $\cos\theta = -\frac{\sqrt{5}}{6}$ for $\frac{\pi}{2} < \theta < \pi$
   $-\frac{\sqrt{155}}{18}, -\frac{13}{18}, \frac{\sqrt{155}}{13}$

4. $\sin\theta = \frac{3\sqrt{6}}{16}$ for $0 < \theta < \frac{\pi}{2}$
   $\frac{3\sqrt{303}}{64}, \frac{37}{64}, \frac{3\sqrt{303}}{37}$

Prove each identity.

5. $\cos 2\theta = \frac{\sin 2\theta - \tan\theta}{\tan\theta}$

   $\cos 2\theta = \frac{2\sin\theta\cos\theta - \tan\theta}{\tan\theta}$;
   $\cos 2\theta = \frac{2\sin\theta\cos\theta - \frac{\sin\theta}{\cos\theta}}{\frac{\sin\theta}{\cos\theta}}$;
   $\cos 2\theta = \frac{2\sin\theta\cos^2\theta - \sin\theta}{\sin\theta}$; $\cos 2\theta =$
   $2\cos^2\theta - 1$; $\cos 2\theta = 2\cos^2\theta - (\sin^2\theta + \cos^2\theta)$;
   $\cos 2\theta = 2\cos^2\theta - \sin^2\theta - \cos^2\theta$;
   $\cos 2\theta = \cos^2\theta - \sin^2\theta$; $\cos 2\theta = \cos 2\theta$

6. $\sin 4\theta = 4\cos\theta\sin\theta - 8\cos\theta\sin^3\theta$

   $\sin 4\theta = 4\cos\theta\sin\theta(1 - 2\sin^2\theta)$;
   $\sin 4\theta = 2\sin 2\theta(1 - 2\sin^2\theta)$;
   $\sin 4\theta = 2\sin 2\theta(\sin^2\theta + \cos^2\theta - 2\sin^2\theta)$
   $\sin 4\theta = 2\sin 2\theta(\cos^2\theta - \sin^2\theta)$;
   $\sin 4\theta = 2\sin 2\theta\cos 2\theta$;
   $\sin 4\theta = \sin 4\theta$

Use half-angle identities to find the exact value of each trigonometric expression.

7. sin 157.5°
   $\frac{\sqrt{2-\sqrt{2}}}{2}$

8. $\cos\frac{5\pi}{12}$
   $\frac{\sqrt{2-\sqrt{3}}}{2}$

9. $\tan\frac{13\pi}{8}$
   $-\sqrt{2} - 1$

Find $\sin\frac{\theta}{2}$, $\cos\frac{\theta}{2}$, and $\tan\frac{\theta}{2}$ for each.

10. $\sin\theta = \frac{1}{5}$ for $90° < \theta < 180°$
    $\sqrt{\frac{5+2\sqrt{6}}{10}}$; $\sqrt{\frac{5-2\sqrt{6}}{10}}$; $\sqrt{\frac{5+2\sqrt{6}}{5-2\sqrt{6}}}$

11. $\cos\theta = \frac{\sqrt{7}}{4}$ for $270° < \theta < 360°$
    $\sqrt{\frac{4-\sqrt{7}}{8}}$; $-\sqrt{\frac{4+\sqrt{7}}{8}}$; $-\sqrt{\frac{4-\sqrt{7}}{4+\sqrt{7}}}$

12. $\tan\theta = 7$ for $180° < \theta < 270°$
    $\sqrt{\frac{10+\sqrt{2}}{20}}$; $-\sqrt{\frac{10-\sqrt{2}}{20}}$; $-\sqrt{\frac{10+\sqrt{2}}{10-\sqrt{2}}}$

13. $\tan\theta = -\frac{\sqrt{5}}{2}$ for $270° < \theta < 360°$
    $\frac{\sqrt{6}}{6}$; $-\frac{\sqrt{30}}{6}$; $-\frac{\sqrt{5}}{5}$

Solve.

14. Amanda wants to shorten the wheelchair ramp at her front door. The current ramp is 13 feet along the horizontal and the step is 9 inches high. If she doubles the measure of the angle that the ramp makes with the walkway, how long will the new ramp be to the nearest inch? 6 ft 6 in.

LESSON 14-5
Reteach
Double-Angle and Half-Angle Identities

There are special double-angle formulas for sin 2θ, cos 2θ, and tan 2θ.

Double-Angle Identities

$\sin 2\theta = 2\sin\theta\cos\theta$

$\cos 2\theta = \cos^2\theta - \sin^2\theta$
$\cos 2\theta = 2\cos^2\theta - 1$
$\cos 2\theta = 1 - 2\sin^2\theta$

$\tan 2\theta = \frac{2\tan\theta}{1 - \tan^2\theta}$

The double angle is twice θ, or 2θ.

There are three equivalent forms of the identity for cos 2θ. If you know one form, you can use $\sin^2\theta + \cos^2\theta = 1$ to derive the other forms.

Find sin 2θ and cos 2θ if $\cos\theta = -\frac{2}{3}$ and $180° < \theta < 270°$.

Notice the interval in which θ lies. Cosine and sine are both negative in Quadrant III.

Step 1 Find sin θ.

$\sin^2\theta = 1 - \cos^2\theta$	*Pythagorean identity*
$\sin^2\theta = 1 - \left(-\frac{2}{3}\right)^2$	*Substitute $\cos\theta = -\frac{2}{3}$.*
$\sin^2\theta = \frac{5}{9}$	*Evaluate.*
$\sin\theta = -\frac{\sqrt{5}}{3}$	*Solve for sin θ.*

Use the negative value since θ lies in Quadrant III.

Step 2 Find sin 2θ.

$\sin 2\theta = 2\sin\theta\cos\theta$	*Write the formula.*
$\sin 2\theta = 2\left(-\frac{\sqrt{5}}{3}\right)\left(-\frac{2}{3}\right)$	*Substitute values for sin θ and cos θ.*
$\sin 2\theta = \frac{4\sqrt{5}}{9}$	*Simplify.*

Step 3 Find cos 2θ.

$\cos 2\theta = 2\cos^2\theta - 1$	*Choose a formula.*
$\cos 2\theta = 2\left(-\frac{2}{3}\right)^2 - 1$	*Substitute the value for cos θ.*
$\cos 2\theta = \frac{8}{9} - 1 = -\frac{1}{9}$	*Simplify.*

Evaluate each expression if $\cos\theta = \frac{1}{4}$ and $270° < \theta < 360°$.

1. sin θ
   Use $\sin^2\theta = 1 - \cos^2\theta$.
   $-\frac{\sqrt{15}}{4}$

2. cos 2θ
   Use $\cos 2\theta = 1 - 2\sin^2\theta$.
   $-\frac{7}{8}$

3. tan θ
   Use $\tan\theta = \frac{\sin\theta}{\cos\theta}$.
   $-\sqrt{15}$

4. tan 2θ
   Use $\tan 2\theta = \frac{2\tan\theta}{1 - \tan^2\theta}$.
   $\frac{\sqrt{15}}{7}$

LESSON 14-5

Reteach

Double-Angle and Half-Angle Identities (continued)

You can use half-angle identities to find the exact value of some trigonometric expressions.

Half-Angle Identities
$\sin\left(\frac{\theta}{2}\right) = \pm\sqrt{\frac{1-\cos\theta}{2}}$: (+) for $\frac{\theta}{2}$ in Quadrants I, II; (−) for $\frac{\theta}{2}$ in Quadrants III, IV
$\cos\left(\frac{\theta}{2}\right) = \pm\sqrt{\frac{1+\cos\theta}{2}}$: (+) for $\frac{\theta}{2}$ in Quadrants I, IV; (−) for $\frac{\theta}{2}$ in Quadrants II, III
$\tan\left(\frac{\theta}{2}\right) = \pm\sqrt{\frac{1-\cos\theta}{1+\cos\theta}}$: (+) for $\frac{\theta}{2}$ in Quadrants I, III; (−) for $\frac{\theta}{2}$ in Quadrants II, IV

Find the exact value of cos 112.5°.

Think: 225° has a reference angle of 45°.

$\cos 112.5° = \cos\frac{225°}{2}$ — *Find the angle: θ = 225°.*

$= -\sqrt{\frac{1+\cos\theta}{2}}$ — *Choose the half-angle identity.*

$= -\sqrt{\frac{1+\cos 225°}{2}}$ — *Substitute θ = 225°.*

112.5° is in QII where cosine is negative.

$= -\sqrt{\frac{1+\left(-\frac{\sqrt{2}}{2}\right)}{2}}$ — *Evaluate.*

$\cos 225° = -\cos 45° = -\frac{\sqrt{2}}{2}$

$= -\sqrt{\left(\frac{2-\sqrt{2}}{2}\right)\left(\frac{1}{2}\right)}$ — *Simplify.*

$= -\sqrt{\frac{2-\sqrt{2}}{4}} = -\frac{\sqrt{2-\sqrt{2}}}{2}$

Use the half-angle identity to find the exact value of each expression.

5. $\tan 157.5° = \tan\frac{315°}{2}$

$\tan\left(\frac{315°}{2}\right) = -\sqrt{\frac{1-\cos 315°}{1+\cos 315°}}$

$-\sqrt{\frac{1-\frac{\sqrt{2}}{2}}{1+\frac{\sqrt{2}}{2}}}$; $-\sqrt{\frac{\frac{2-\sqrt{2}}{2}}{\frac{2+\sqrt{2}}{2}}}$;

$-\sqrt{\frac{2-\sqrt{2}}{2+\sqrt{2}}}$

6. $\sin 15° = \sin\frac{30°}{2}$

$\sin\left(\frac{30°}{2}\right) = \sqrt{\frac{1-\cos 30°}{2}}$;

$\sqrt{\frac{1-\frac{\sqrt{3}}{2}}{2}}$;

$\sqrt{\frac{2-\sqrt{3}}{4}}$; $\frac{\sqrt{2-\sqrt{3}}}{2}$

LESSON 14-5

Challenge

Double-Angle and Half-Angle Identities for Tangent

Double-angle and half-angle identities for sine and cosine are shown below.

Double-Angle Identities	*Half-Angle Identities*
$\sin 2A = 2\sin A\cos A$	$\sin\frac{A}{2} = \pm\sqrt{\frac{1-\cos A}{2}}$
$\cos 2A = \cos^2 A - \sin^2 A$	$\cos\frac{A}{2} = \pm\sqrt{\frac{1+\cos A}{2}}$

You can use these identities to derive double-angle and half-angle identities for tangent.

1. a. Use a ratio identity to express tan2*A* in terms of sin2*A* and cos2*A*. $\tan 2A = \frac{\sin 2A}{\cos 2A}$
   b. Use identities for sin2*A* and cos2*A* to rewrite the right side in terms of sin*A* and cos*A*. $\tan 2A = \frac{2\sin A\cos A}{\cos^2 A - \sin^2 A}$
   c. Divide numerator and denominator of the right side by $\cos^2 A$. $\tan 2A = \frac{\frac{2\sin A\cos A}{\cos^2 A}}{\frac{\cos^2 A}{\cos^2 A} - \frac{\sin^2 A}{\cos^2 A}}$
   d. Simplify, and rewrite the right side in terms of tan*A*. $\tan 2A = \frac{2\tan A}{1-\tan^2 A}$
2. Verify the result of Exercise 1 by using the identity shown below to obtain an identity for tan2*A*.
   $\tan(A+B) = \frac{\tan A + \tan B}{1 - \tan A\tan B}$
   $\tan 2A = \tan(A+A) = \frac{\tan A + \tan A}{1 - \tan A\tan A} = \frac{2\tan A}{1-\tan^2 A}$
3. Angle *A* is in Quadrant II and $\sin A = \frac{4}{5}$.
   a. Use your identity to determine tan2*A*. $\tan 2A = \frac{24}{7} \approx 3.43$
   b. Use a calculator to verify your result. $m\angle A = 127°$, $\tan 254° \approx 3.49$
4. Show that $2\cot 2A = \cot A - \tan A$ is an identity by
   a. expressing 2cot2*A* in terms of tan*A*. $2\cot 2A = \frac{1-\tan^2 A}{\tan A}$
   b. expressing cot*A* − tan*A* as a single fraction in terms of tan*A*. $\cot A - \tan A = \frac{1-\tan^2 A}{\tan A}$
5. Derive an identity for $\tan\frac{A}{2}$ in terms of cos*A*. $\tan\frac{A}{2} = \pm\sqrt{\frac{1-\cos A}{1+\cos A}}$
6. a. Use your identity to determine tan22.5°. $\tan 22.5° = \sqrt{\frac{2-\sqrt{2}}{2+\sqrt{2}}}$
   b. Verify that your result is equivalent to $\sqrt{2} - 1$. $\sqrt{\frac{2-\sqrt{2}}{2+\sqrt{2}}\cdot\frac{2-\sqrt{2}}{2-\sqrt{2}}} = \frac{2-\sqrt{2}}{\sqrt{2}}\cdot\frac{\sqrt{2}}{\sqrt{2}} = \sqrt{2}-1$
7. Show that $\tan\frac{A}{2}$ is equivalent to $\frac{1-\cos A}{\sin A}$.
   $\tan\frac{A}{2} = \sqrt{\frac{1-\cos A}{1+\cos A}\cdot\frac{1-\cos A}{1-\cos A}} = \sqrt{\frac{(1-\cos A)^2}{\sin^2 A}} = \frac{1-\cos A}{\sin A}$

LESSON 14-5

Problem Solving

Double-Angle and Half-Angle Identities

Chris set a new personal best in football with a 35-yard kick. He wants to improve his distance. His coach suggests using the formula $d(\theta) = \frac{3v_0^2\sin\theta\cos\theta}{16}$, which gives the horizontal distance, *d*, in yards, as a function of the angle of elevation, θ, at which the ball is kicked with an initial velocity of v_0 yards per second.

1. What is the angle of elevation of the ball when Chris kicks it with an initial velocity of 20 yards per second and the ball covers 35 yards.
   a. What double-angle identity can you use to rewrite the function in terms of sin2θ? $\sin 2\theta = 2\sin\theta\cos\theta$
   b. Rewrite the function in terms of sin2θ. $d(\theta) = \frac{3v_0^2\sin 2\theta}{32}$
   c. Substitute the known values in the equation. $35 = \frac{3(20)^2\sin 2\theta}{32}$
   d. Solve for sin2θ. $\sin 2\theta = 0.933$
   e. What is the angle at which he kicked the ball? 34.5°
2. Chris wonders how the horizontal distance would change if he kicked the ball with the same velocity but at a different angle.
   a. Complete the table to show the horizontal distance, to the nearest hundredth yard, for the given angles.
   b. At what angle must Chris kick the ball to cover the greatest horizontal distance? 45°

Angle of Elevation (°)	Horizontal Distance (yd)
15	18.75
25	28.73
35	35.24
45	37.50
55	35.24

3. What is the least angle at which Chris must kick the ball if he wants to cover at least 30 yards horizontally? 26.5°
4. Chris thinks he can kick the ball with an initial velocity of 18 yards per second and make a horizontal distance of 40 yards. Is he correct? Explain.
   Possible answer: For this initial velocity and target distance, $\sin 2\theta \approx 1.32$, which is outside the possible range of values for the sine of an angle; so he is incorrect.

Choose the letter for the best answer.

5. If Chris kicks the ball with an initial velocity of only 15 yards per second and an angle of elevation of 45°, what is the maximum horizontal distance the ball will cover?
   A 12.06 yd
   B 14.92 yd
   (C) 21.09 yd
   D 29.83 yd
6. At what angle can Chris kick the ball to cover a horizontal distance of 36 yards, if he kicks the ball with an initial velocity of 20 yards per second?
   F 18.5°
   (G) 36.9°
   H 44.1°
   J 73.7°

LESSON 14-5

Reading Strategy

Analyze Information

You can use double-angle identities and half-angle identities to evaluate expressions and find exact values. One step is to find the sign of the functions of the angle.

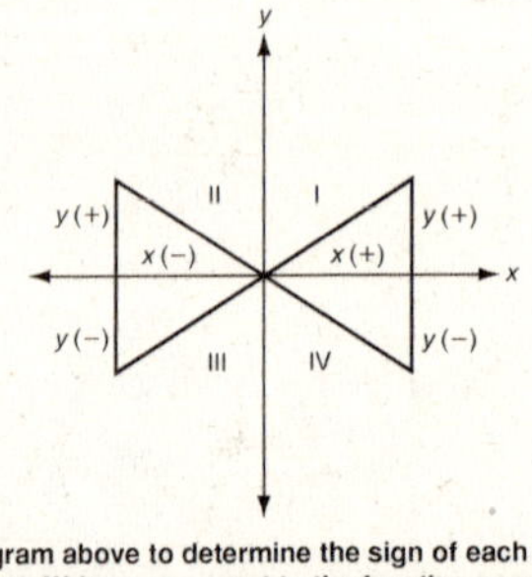

Use the diagram above to determine the sign of each function of θ in each quadrant. Write + or − next to the function name.

	Quadrant I	Quadrant II	Quadrant III	Quadrant IV
1.	sin +	sin +	sin −	sin −
2.	cos +	cos −	cos −	cos +
3.	tan +	tan −	tan +	tan −

Determine the quadrant and sign of each trigonometric expression.

4. sin330° QIV, −
5. cos210° QIII, −
6. sin150° QII, +
7. tan123° QII, −
8. sin290° QIV, −
9. cos13° QI, +
10. sin194° QIII, −
11. cos359° QIV, +
12. tan275° QIV, −
13. cos178° QII, −
14. tan205° QIII, +
15. sin85° QI, +

LESSON 14-6

Practice A

Solving Trigonometric Equations

Find all of the solutions of each equation.

1. $3\sin\theta = \sin\theta - \sqrt{3}$

   a. Subtract $\sin\theta$ from both sides and combine like terms. $2\sin\theta = -\sqrt{3}$

   b. Solve for $\sin\theta$. $\sin\theta = -\frac{\sqrt{3}}{2}$

   c. Apply the inverse sine. $240° + 360°n,\ 300° + 360°n$

2. $5\cos\theta = 3\cos\theta - 1$ $120° + 360°n,\ 240° + 360°n$

3. $10\tan\theta = 9\tan\theta + 1$ $45° + 180°n$

Solve each equation for the given domain.

4. $\cos^2\theta - 2\cos\theta = -1$ for $0° \le \theta \le 360°$

   a. Add 1 to both sides. $\cos^2\theta - 2\cos\theta + 1 = 0$

   b. Factor or use the quadratic formula to solve for $\cos\theta$. $(\cos\theta - 1)^2 = 0$

   c. Apply the inverse cosine. $0°, 360°$

5. $\tan^2\theta = 9$ for $0° \le \theta \le 360°$ $\approx 71.6°, \approx 108.4°, \approx 251.6°, \approx 288.4°$

6. $4\sin^2\theta = 3\sin\theta$ for $0° \le \theta \le 360°$ $0°, 180°, 360°, 48.6°, 138.6°$

Solve each equation for the given domain. Use trigonometric identities.

7. $\cos^2\theta = \sin^2\theta + \sin\theta$ for $0° \le \theta \le 360°$

   a. Rewrite the equation in terms of $\sin\theta$. $1 - \sin^2\theta = \sin^2\theta + \sin\theta$

   b. Solve for $\sin\theta$. $\sin\theta = -1,\ \sin\theta = \frac{1}{2}$

   c. Apply the inverse sine. $30°, 150°, 270°$

8. $\cos 2\theta + \sin\theta = 1$ for $0° \le \theta \le 360°$ $0°, 30°, 150°, 180°, 360°$

9. $\sin 2\theta \tan\theta = 1$ for $0° \le \theta \le 360°$ $45°, 135°, 225°, 315°$

Solve.

10. The horizontal range, R, in feet that a baseball travels is modeled by $R(\theta) = \frac{v^2 \sin 2\theta}{32}$. The initial velocity, v, is 90 feet per second. The ball is pitched at an angle of θ degrees with the horizontal. At what angle(s) will the ball travel 200 feet? $26°, 64°$

LESSON 14-6

Practice B

Solving Trigonometric Equations

Find all of the solutions of each equation.

1. $4\tan\theta = 5\tan\theta + \sqrt{3}$ $120° + 180°n$

2. $\sqrt{2} - 2\sin\theta = 0$ $45° + 360°n,\ 135° + 360°n$

3. $7\cos\theta - 1 = 9\cos\theta$ $120° + 360°n,\ 240° + 360°n$

4. $2\tan\theta + \sqrt{3} = 5\tan\theta$ $30° + 180°n$

Solve each equation for the given domain.

5. $\tan^2\theta - 2\tan\theta = -1$ for $0° \le \theta \le 360°$ $45°, 225°$

6. $2\sin^2\theta = 1$ for $0° \le \theta \le 360°$ $45°, 135°, 225°, 315°$

7. $4\cos^2\theta = 3\cos\theta$ for $90° \le \theta \le 180°$ $90°$

8. $2\sqrt{3}\cos^2\theta - \cos\theta = 2\sqrt{3}$ for $\pi \le \theta \le 2\pi$ $\frac{7\pi}{6}$

9. $3\tan^2\theta = 2\tan\theta + 1$ for $0° \le \theta \le 90°$ $45°$

10. $5\tan^2\theta = 2\tan\theta + 6$ for $180° \le \theta \le 360°$ $232.7°, 317.6°$

Use trigonometric identities to solve each equation for the given domain.

11. $\sin\theta + \sin 2\theta = 0$ for $0° \le \theta \le 360°$ $0°, 120°, 180°, 240°, 360°$

12. $\cos^2\theta = \sin^2\theta$ for $0° \le \theta \le 360°$ $45°, 135°, 225°, 315°$

13. $\frac{\cos 2\theta}{\sin^2\theta} = 0$ for $0° \le \theta \le 360°$ $45°, 135°, 225°, 315°$

14. $\cos 2\theta = \frac{1}{2}\sin\theta$ for $0° \le \theta \le 90°$ $\approx 36.4°, \approx 57.5°$

Solve.

15. The height of the water at a pier on a certain day can be modeled by $h(t) = 4.8\sin\frac{\pi}{6}(t + 3.5) + 9$, where h is the height in feet and t is the time in hours after midnight. When is the height of the water 6 feet? 3:47 A.M., 7:13 A.M., 3:47 P.M., 7:13 P.M.

LESSON 14-6

Practice C

Solving Trigonometric Equations

Find all of the solutions of each equation.

1. $4\tan\theta + \sqrt{3} = \tan\theta$ $150° + 180°n$

2. $\cos\theta = 1 + 3\cos\theta$ $120° + 360°n,\ 240° + 360°n$

Solve each equation for the given domain.

3. $4\cos^2\theta = 1$ for $0° \le \theta \le 360°$ $60°, 120°, 240°, 300°$

4. $\cos^2\theta + \cos\theta - 1 = 0$ for $0° \le \theta \le 180°$ $\approx 51.8°$

5. $3\tan^2\theta + \sqrt{3} = 3\tan\theta + \sqrt{3}\tan\theta$ for $\pi \le \theta \le 2\pi$ $\frac{7\pi}{6}, \frac{5\pi}{4}$

Use trigonometric identities to solve each equation for the given domain.

6. $\sin 2\theta = \sqrt{2}\cos\theta$ for $0° \le \theta \le 360°$ $45°, 90°, 135°, 225°, 270°, 315°$

7. $\cos\frac{\theta}{2} = \sin\theta$ for $0° \le \theta \le 360°$ $60°, 180°, 300°$

8. $\tan 2\theta = \cot\theta$ for $0° \le \theta \le 180°$ $30°, 90°, 150°$

9. $2\cos^2\frac{\theta}{2} = \cos 2\theta$ for $0° \le \theta \le 360°$ $\approx 141.3°, \approx 218.7°$

10. $\sin\left(\frac{\pi}{4} - \theta\right) = 1 - \sin\left(\frac{\pi}{4} + \theta\right)$ for $0 \le \theta \le 2\pi$ $\frac{\pi}{4}, \frac{7\pi}{4}$

11. $\cos 2\theta \tan\theta = -\sin\theta$ for $0° \le \theta \le 360°$ $0°, 60°, 180°, 300°, 360°$

12. $\cos\theta + \sin\theta = 1$ for $0° \le \theta \le 360°$ $0°, 90°, 360°$

13. $\cos^3\theta - \sin^3\theta = 1$ for $0° \le \theta \le 360°$ $0°, 270°, 360°$

Solve.

14. The number of hours of daylight, d, at Quintilla Falls can be modeled by $h(d) = 3.22\sin\frac{\pi}{182.5}(d - 88) + 12.32$, where d is the number of days after January 1. On what dates are there 12 hours of sunlight? Assume the year is not a leap year. March 24 and October 4

LESSON 14-6

Reteach

Solving Trigonometric Equations

You can use the same methods to solve trigonometric equations as to solve algebraic equations.

Substitute variables to solve trigonometric equations that resemble quadratic equations.

Solve: $\sin^2\theta - 3\sin\theta - 4 = 0$ for $0° \le \theta < 360°$

The interval includes 0° but not 360°.

Step 1 Make a substitution.
Let $x = \sin\theta$ in $\sin^2\theta - 3\sin\theta - 4 = 0$
$x^2 - 3x - 4 = 0$

Step 2 Solve the quadratic $x^2 - 3x - 4 = 0$.
$(x - 4)(x + 1) = 0$ *Factor.*
$x - 4 = 0$ $x + 1 = 0$ *Set each factor equal to 0.*
$x = 4$ $x = -1$ *Solve each equation for x.*

Step 3 Substitute $\sin\theta$ for x.
$\sin\theta = 4$ *This equation has no solution.* ($-1 \le \sin\theta \le 1$)
$\sin\theta = -1$ *Think: What angle has a sine of −1?*
$\theta = 270°$

The only solution is $\theta = 270°$.

Complete to solve each equation for $0° \le \theta < 360°$.

1. $3\tan^2\theta - 1 = 0$

   Let $x = \tan\theta$.

   Hint: There are 4 solutions.

   $3x^2 - 1 = 0$

   $3x^2 = 1$

   $x = \pm\frac{\sqrt{3}}{3}$

   $\tan\theta = \frac{\sqrt{3}}{3}$ or $-\frac{\sqrt{3}}{3}$,

   $30°, 150°, 210°, 330°$

2. $2\sin^2\theta - \sin\theta = 0$

   Let $x = \sin\theta$.

   Hint: There are 4 solutions.

   $2x^2 - x = 0$

   $x(2x - 1) = 0$

   $x = 0$ or $2x - 1 = 0$

   $x = 0$ or $x = \frac{1}{2}$

   $\sin\theta = 0$ or $\sin\theta = \frac{1}{2}$

   $0°, 30°, 150°, 180°$

3. $\cos^2\theta - 2\cos\theta + 1 = 0$

   $x^2 - 2x + 1;\ (x - 1)^2;$

   $\cos\theta = 1;\ 0°$

4. $2\cos^2\theta - 7\cos\theta + 3 = 0$

   $2x^2 - 7x + 3 = 0$

   $(2x - 1)(x - 3) = 0$

   $x = \frac{1}{2}$ or $x = 3$

   $60°, 300°$

LESSON **14-6**

Reteach

Solving Trigonometric Equations (continued)

You can use trigonometric identities to solve trigonometric equations. This is especially useful when more than one trigonometric function appears in an equation.

> Use an identity to rewrite the equation using a single trigonometric function or a single angle.

Solve: $\cos 2\theta + \cos\theta = 0$ for $0° \le \theta < 360°$

> Use an identity. Choose the formula that allows you to write the equation in terms of $\cos\theta$ only.

Step 1 The angles are not the same.
Use the double-angle identity for cosine.
Choose $\cos 2\theta = 2\cos^2\theta - 1$.

Step 2 Substitute $2\cos^2\theta - 1$ for $\cos 2\theta$ in the original equation.

$\cos 2\theta + \cos\theta = 0$	*Write the original equation.*
$(2\cos^2\theta - 1) + \cos\theta = 0$	*Substitute.*
$2\cos^2\theta + \cos\theta - 1 = 0$	*Write in standard form.*

Step 3 Solve the equation.

$2\cos^2\theta + \cos\theta - 1 = 0$		
$(2\cos\theta - 1)(\cos\theta + 1) = 0$		*Factor.*
$2\cos\theta - 1 = 0$	$\cos\theta + 1 = 0$	*Set each factor equal to 0.*
$\cos\theta = \frac{1}{2}$	$\cos\theta = -1$	*Solve each equation for $\cos\theta$.*

Step 4 Find the values that satisfy each solution.

$\cos\theta = \frac{1}{2}$ Think: What angle has a cosine of $\frac{1}{2}$?
$\theta = 60°, 300°$
$\cos\theta = -1$ Think: What angle has a cosine of -1?
$\theta = 180°$

> Remember to find all values in the interval.

Use trigonometric identities to solve each equation for $0° \le \theta < 360°$.

5. $\cos 2\theta + \sin\theta + 2 = 0$

Use $\cos 2\theta = 1 - 2\sin^2\theta$.

$1 - 2\sin^2\theta + \sin\theta + 2 = 0$
$2\sin^2\theta - \sin\theta - 3 = 0$
$(2\sin\theta - 3)(\sin\theta + 1) = 0$
$\sin\theta = \frac{3}{2}$ or $\sin\theta = -1$
270°

6. $\cos^2\theta + 2\sin\theta = 1$

Use $\cos^2\theta = 1 - \sin^2\theta$.

$1 - \sin^2\theta + 2\sin\theta = 1$
$\sin^2\theta - 2\sin\theta = 0$
$\sin\theta(\sin\theta - 2) = 0$
$\sin\theta = 0$ or $\sin\theta = 2$
0°, 180°

LESSON **14-6**

Challenge

Trigonometric Equations with Radicals

A *trigonometric equation* contains one or more trigonometric functions of an angle. Such an equation must first be solved for a function of the angle and then for the measure of the angle. Some trigonometric equations contain radicals.

1. Solve $2\cos\theta = \sqrt{8\sin\theta - 1}$ for $0° \le \theta < 360°$.
   a. Since the radical is isolated, square both sides of the equation. $4\cos^2\theta = 8\sin\theta - 1$
   b. Use an identity to express the equation in terms of a single trigonometric function. $4(1 - \sin^2\theta) = 8\sin\theta - 1$
   c. Solve for θ and check. 30°

Solve each equation for $0° \le \theta < 360°$. Check your results.

2. $\sqrt{\frac{\sin\theta}{2}} = \sin\theta$ 0°, 30°, 150°, 180°

3. $\sqrt{2 + 3\cot^2\theta} = \tan\theta$ 60°, 240°

4. $\sin\theta + \sin\frac{\theta}{2} = 0$
   a. Use an identity to replace $\sin\frac{\theta}{2}$. $\sin\theta \pm \sqrt{\frac{1 - \cos\theta}{2}} = 0$
   b. Isolate the radical and square both sides. $\sin^2\theta = \frac{1 - \cos\theta}{2}$
   c. Use an identity to express the equation in terms of a single trigonometric function. $1 - \cos^2\theta = \frac{1 - \cos\theta}{2}$
   d. Solve for θ and check. 0°, 240°

5. $\cos\theta = \sin\frac{\theta}{2}$ 60°, 300°

6. $\sin\theta = 2\sin\frac{\theta}{2}$ 0°

As opposed to the equations in Exercises 4–6, where you first had to get the angles to be of the same multiple of θ, the angles in Exercises 7 and 8 are already of the same multiple. So, you do not need to use radicals.

Solve each equation for $0° \le \theta < 360°$.

7. $\cos\frac{\theta}{2} + \sin\frac{\theta}{2} = 1$ 0°, 180°

8. $2\sin\theta\cos\theta + 2\sin\theta - \cos\theta - 1 = 0$ 30°, 150°, 180°

LESSON **14-6**

Problem Solving

Solving Trigonometric Equations

Jon watches his sister Jessica skateboard. When she jumps and launches herself into the air with an initial speed of v_0 feet per second, her path in terms of time, t, in seconds, is represented by these equations.

Equation 1 $x(t) = v_0 t\cos\theta$

Equation 2 $y(t) = v_0 t\sin\theta - 16t^2$

The first equation models the horizontal distance, $x(t)$, that the skateboarder travels, and the second equation models the vertical height, $y(t)$, that the skateboarder attains.

1. Jessica attains a height of 4.7 feet above the launch and landing ramps after 1 second. Her initial velocity is 25 feet per second. Find the angle of her launch.
   a. Which equation can you use with the given information to solve for θ? Equation 2
   b. Substitute the known values and solve for θ. $\theta = 55.9°$
   c. What is Jessica's height above the launch and landing ramps after 0.5 second? 6.4 ft
   d. What distance has Jessica traveled after 1 second? 14 ft
2. Jon attains a height of 5.2 feet above the launch and landing ramps after 1 second. His initial velocity is 28 feet per second. Find the angle of his launch.
   a. Write and evaluate an expression for Jon's launch angle.
   $5.2 = 28(1)\sin\theta - 16(1)^2$; $\theta = 49.2°$
   b. Write and evaluate an expression for his height above the launch and landing ramps after 0.5 second.
   $y(t) = 28(0.5)\sin 49.2° - 16(0.5)^2$; $y(t) = 6.6$ ft

Choose the letter for the best answer.

3. How far does Jon travel in 1 second?
   A 14.5 ft
   (C) 18.3 ft
   B 16.7 ft
   D 21.2 ft
4. Jessica increases her initial velocity to 30 feet per second. She attains a height of 5.5 feet after 1 second. Which expression represents her launch angle?
   F $\cos^{-1}\left(\frac{10.5}{30}\right)$
   H $\sin^{-1}\left(\frac{10.5}{30}\right)$
   G $\cos^{-1}\left(\frac{21.5}{30}\right)$
   (J) $\sin^{-1}\left(\frac{21.5}{30}\right)$

LESSON **14-6**

Reading Strategy

Compare and Contrast

You can solve trigonometric equations in the same way as you solve algebraic equations.

Linear Algebraic Equations	Linear Trigonometric Equations
$3x = x + 2$ $3x - x = 2$ $2x = 2$ $x = 1$	$3\tan\theta = \tan\theta + 2$ $3\tan\theta - \tan\theta = 2$ $2\tan\theta = 2$ $\tan\theta = 1$
Quadratic Algebraic Equations	**Quadratic Trigonometric Equations**
$x^2 + 3x - 10 = 0$ $(x + 5)(x - 2) = 0$ $x + 5 = 0$ or $x - 2 = 0$ $x = -5$ or $x = 2$	$2\cos^2\theta - \cos\theta - 1 = 0$ $(2\cos\theta + 1)(\cos\theta - 1) = 0$ $2\cos\theta + 1 = 0$ or $\cos\theta - 1 = 0$ $\cos\theta = -\frac{1}{2}$ or $\cos\theta = 1$

Solve.

1. $3\cot\theta = \cot\theta + 1$ $\cot\theta = \frac{1}{2}$
2. $2\sec\theta = 6\sec\theta - 3$ $\sec\theta = \frac{3}{4}$
3. $\frac{5}{2}\cos\theta + \frac{3}{2}\cos\theta - 4 = 0$ $\cos\theta = 1$
4. $5\tan\theta + 3 = 4\tan\theta - \frac{1}{3}$ $\tan\theta = -\frac{10}{3}$
5. $6\sin\theta = 3\sin\theta + 3$ $\sin\theta = 1$
6. $\frac{1}{2}\cot\theta = -\frac{5}{2}\cot\theta + 2$ $\cot\theta = \frac{2}{3}$
7. $3\cos^2\theta - 1 = 2\cos^2\theta - \frac{3}{4}$ $\cos\theta = \pm\frac{1}{2}$
8. $\sin^2\theta + 3\sin\theta + 5 = \sin\theta + 4 + \sin^2\theta$ $\sin\theta = -\frac{1}{2}$
9. $3\tan^2\theta = \tan\theta$ $\tan\theta = 0$ or $\frac{1}{3}$
10. $4\cos^2\theta = 1 - 3\cos\theta$ $\cos\theta = -1$ or $\frac{1}{4}$

Holt Algebra 2

Chapter 11 Resource Book

HOLT, RINEHART AND WINSTON
A Harcourt Education Company
Orlando • Austin • New York • San Diego • London

Printed in the United States of America

ISBN 978-0-03-042819-7

ISBN 0-03-042819-X

5 6 7 8 9 862 09 08

Contents

Date ______________

Dear Family,

Chapter 11 introduces your child to several topics from the fields of probability and statistics.

When you have several groups of items and you select only one item from each group, the **Fundamental Counting Principle** allows you to multiply the number of items in each group to find the total number of selections. For example, if you have 5 shirts, 3 pants, and 4 jackets, you can create 60 outfits because $5 \cdot 3 \cdot 4 = 60$.

A **permutation** is a selection of items from one group in which order is important. A **combination** is a selection in which order is *not* important. Calculating permutations and combinations require the use of **factorial** (!), the product of the natural numbers less than or equal to the number.

Type	Formula	Example
Permutation	${}_nP_r = \dfrac{n!}{(n-r)!}$	How many ways can seven swimmers finish in first, second, and third place? ${}_7P_3 = \dfrac{7!}{(7-3)!} = \dfrac{7!}{4!} = \dfrac{7\cdot 6\cdot 5\cdot \cancel{4\cdot 3\cdot 2\cdot 1}}{\cancel{4\cdot 3\cdot 2\cdot 1}} = 210$
Combination	${}_nC_r = \dfrac{n!}{r!(n-r)!}$	How many ways can seven coworkers be chosen for a committee of three? ${}_7C_3 = \dfrac{7!}{3!(7-3)!} = \dfrac{7!}{3!(4!)} = \dfrac{7\cdot \cancel{6}\cdot 5\cdot \cancel{4\cdot 3\cdot 2\cdot 1}}{\cancel{3\cdot 2}\cdot 1\cdot (\cancel{4\cdot 3\cdot 2\cdot 1})} = 35$

Probability is a measure of how likely an event is to occur. To find a **theoretical probability**, you assume all events are equally likely and calculate the ratio of the number of favorable outcomes to the number of all possible outcomes.

To find an **experimental probability**, you use real-world data and calculate the ratio of the number of times an event actually occurs to the number of trials of the experiment.

Probability	Formula	Example
Theoretical	$P(\text{event}) = \dfrac{\text{number of favorable outcomes}}{\text{number of all possible outcomes}}$	What is the theoretical probability of rolling a 3 on a number cube? $P(3) = \dfrac{1}{6}$ ← only one 3 on the cube; ← 6 total sides on the cube
Experimental	$P(\text{event}) = \dfrac{\text{number of times the event occurs}}{\text{number of trials}}$	A number cube is rolled one-hundred times and comes up 3 forty times. What is the experimental probability? $P(3) = \dfrac{40}{100} = \dfrac{2}{5}$ ← 40 favorable rolls; ← 100 total rolls

Events can be **independent** (one event does *not* influence the other), **dependent** (one event *does* influence the other), **mutually exclusive** (the events *cannot* both occur during the same trial), or **inclusive** (the events *can* occur at the same time).

Situation	Formula	Example
Independent Events	$P(A \text{ and } B) = P(A) \cdot P(B)$	Find the probability of rolling a 3 on one number cube and a 3 on another number cube. $P(3 \text{ and } 3) = \frac{1}{6} \cdot \frac{1}{6} = \frac{1}{36}$
Dependent Events	$P(A \text{ and } B) = P(A) \cdot P(B \text{ given } A)$	Find the probability of drawing two aces from a deck of cards without replacement. $P(\text{A and A}) = \frac{4}{52} \cdot \frac{3}{51} = \frac{12}{2652}$
Mutually Exclusive Events	$P(A \text{ or } B) = P(A) + P(B)$	Find the probability of rolling a 3 or a 4 on one number cube. $P(3 \text{ or } 4) = \frac{1}{6} + \frac{1}{6} = \frac{2}{6}$
Inclusive Events	$P(A \text{ or } B) = P(A) + P(B) - P(A \text{ and } B)$	Find the probability of drawing an ace or a red card from a deck of cards. $P(\text{A or red}) = \frac{4}{52} + \frac{26}{52} - \frac{2}{52} = \frac{28}{52}$

You and your child may already be familiar with the statistical measures of central tendency: **mean** (the average), **median** (the middle), and **mode** (the most often). Chapter 11 also uses measures of variation, including the **range** (the difference between the maximum and minimum), the **interquartile range** (the "length" of the middle 50% of data), and the **standard deviation** (a complex average of how much each data value differs from the mean).

When a binomial of the form $(p + q)$ is raised to a power and expanded, the terms of the polynomial are given by the **Binomial Theorem**:

$$(p+q)^n = {}_nC_0 p^n q^0 + {}_nC_1 p^{n-1} q^1 + {}_nC_2 p^{n-2} q^2 + \cdots + {}_nC_{n-1} p^1 q^{n-1} + {}_nC_n p^0 q^n$$

Surprisingly, the terms also correspond to the probabilities of events in a situation in which each trial has two possible outcomes, success or failure. If p and q are the probabilities of success or failure, respectively, then the **binomial probability** of r successes is $P(r) = {}_nC_r p^r q^{n-r}$. For example, if each item on a multiple-choice test has four options, then the probability of guessing 8 out of 10 correctly is $P(8) = {}_{10}C_8 \left(\frac{1}{4}\right)^8 \left(\frac{3}{4}\right)^2 = \frac{405}{1{,}048{,}576} \approx 0.0004$.

For additional resources, visit go.hrw.com and enter the keyword MB7 Parent.

Name ______________________ Date ____________ Class ____________

LESSON 11-1

Practice A

Permutations and Combinations

Use the Fundamental Counting Principle.

1. For her aquarium, Susan can choose from 4 types of fish and 3 types of plants. If she chooses one type of fish and one type of plant, how many different aquariums can Susan set up? ____________

2. Lottery numbers in a particular state consist of 6 digits. Each lottery ball that is drawn has six sides, numbered 1–6. How many different lottery numbers are possible? ____________

Evaluate.

3. 5! ____________

4. $\frac{6!}{2!}$ ____________

5. $\frac{2!3!}{4!}$ ____________

Solve.

6. For an art exhibit, Craig has to choose 3 ceramic mugs out of the 7 that he made over the summer. In how many ways can he arrange these 3 mugs in a row? ____________

7. The members of a track team want to choose a team captain and someone to organize their equipment. In how many ways can 2 people be chosen from a team of 10 girls? ____________

Evaluate.

8. ${}_4P_3$ ____________

9. ${}_3C_2$ ____________

10. ${}_5P_2$ ____________

11. ${}_5C_2$ ____________

12. ${}_8P_4$ ____________

13. ${}_6C_1$ ____________

Solve.

14. While on vacation, Sandra wants to buy 2 wallets. There are 7 different patterns she can choose from. In how many ways can Sandra choose 2 different wallets? ____________

15. Vince chooses 3 side dishes from a total of 10 side dishes offered on the menu. In how many different ways can he choose his side dishes? ____________

Name ______________________ Date ____________ Class ____________

LESSON 11-1

Practice B

Permutations and Combinations

Use the Fundamental Counting Principle.

1. The soccer team is silk-screening T-shirts. They have 4 different colors of T-shirts and 2 different colors of ink. How many different T-shirts can be made using one ink color on a T-shirt? ____________

2. A travel agent is offering a vacation package. Participants choose the type of tour, a meal plan, and a hotel class from the table below.

Tour	Meal	Hotel
Walking	Restaurant	4-Star
Boat	Picnic	3-Star
Bicycle		2-Star
		1-Star

How many different vacation packages are offered? ____________

Evaluate.

3. $\frac{3!6!}{3!}$ ____________

4. $\frac{10!}{7!}$ ____________

5. $\frac{9! - 6!}{(9 - 6)!}$ ____________

Solve.

6. In how many ways can the debate team choose a president and a secretary if there are 10 people on the team? ____________

7. A teacher is passing out first-, second-, and third-place prizes for the best student actor in a production of *Hamlet*. If there are 14 students in the class, in how many different ways can the awards be presented? ____________

Evaluate.

8. ${}_5P_4$ ____________

9. ${}_3C_2$ ____________

10. ${}_8P_3$ ____________

Solve.

11. Mrs. Marshall has 11 boys and 14 girls in her kindergarten class this year.

a. In how many ways can she select 2 girls to pass out a snack? ____________

b. In how many ways can she select 5 boys to pass out new books? ____________

c. In how many ways can she select 3 students to carry papers to the office? ____________

Name ________________________________ Date ______________ Class ______________

LESSON 11-1

Practice C

Permutations and Combinations

Evaluate.

1. $\frac{7! - 4!}{(6 - 3)!}$ ______________

2. $\frac{6!}{3!(8 - 5)!}$ ______________

3. $\frac{5!4!}{9!}$ ______________

4. ${}_{10}C_5$ ______________

5. ${}_{7}P_4$ ______________

6. ${}_{10}C_9$ ______________

Compare. Write >, <, or =.

7. ${}_{8}P_3 \,\square\, {}_{6}C_3$ ______________

8. ${}_{12}C_9 \,\square\, {}_{12}C_7$ ______________

9. ${}_{9}P_5 \,\square\, {}_{10}P_5$ ______________

Solve.

10. The door code to get into a top-secret laboratory is 6 digits. The first 3 digits of the code are all odd and the last 3 digits are all even. Digits can be used more than once. How many possible codes are there to gain access to this laboratory? ______________

11. In how many ways can a 3-digit number be formed using the numbers 0–9, if each digit is used only one time? ______________

12. The principal of the high school selects 4 Merit Scholars to attend a Town Council meeting. If there are a total of 12 Merit Scholars at the school, in how many ways can the students be selected? ______________

13. A board of trustees is made up of 10 people. The board is choosing a chairperson, a secretary, and a publicist. If they have already decided upon a chairperson, in how many ways can they choose a secretary and a publicist? ______________

14. There are 8 marbles in a bag. If they are all different colors, in how many ways can 4 marbles be chosen? ______________

15. A student in a biology laboratory has 7 plants to use in an experiment. One plant will act as the control, 3 will be subjected to Environment A, and 3 will be subjected to Environment B. In how many ways can the student choose the plants that will be subjected to Environment B? ______________

16. Holly wants to choose 5 different decorative tiles out of 8. If she plans to place the 5 tiles in a row, end to end, in how many different ways can she arrange them, from left to right? ______________

Name ______________________ Date __________ Class __________

LESSON 11-1

Reteach

Permutations and Combinations

A **permutation** is a selection of items from a group in which the order is important. In a permutation, *AB* is NOT the same as *BA*.

The number of permutations of *n* items taken *r* at a time is shown by the following formula.

$$_nP_r = \frac{n!}{(n-r)!}$$

The value of *r* must be less than or equal to the value of *n*.

How many ways can club members select a president, a vice president, a secretary, and a treasurer from a group of 10 members?
Order matters since each office is different.

To find the number of permutations of 10 items taken 4 at a time, use $n = 10$ and $r = 4$ in the permutation rule. Then evaluate.

$$_{10}P_4 = \frac{10!}{(10-4)!} = \frac{10!}{6!} = \frac{10 \cdot 9 \cdot 8 \cdot 7 \cdot \cancel{6 \cdot 5 \cdot 4 \cdot 3 \cdot 2 \cdot 1}}{\cancel{6 \cdot 5 \cdot 4 \cdot 3 \cdot 2 \cdot 1}} = 10 \cdot 9 \cdot 8 \cdot 7 = 5040$$

Remember that *n*! or "*n* factorial" means to find the product of the whole numbers from 1 to *n*.

There are 5040 ways to select the officers.

Evaluate.

1. 8! ____________

2. 5! ____________

3. 10! ____________

4. $\frac{6!}{3!}$ ____________

5. $\frac{9!}{4!}$ ____________

6. $\frac{15!}{14!}$ ____________

Solve.

7. How many ways can the letters from *A* through *H* be used to create 5-letter passwords is there are no repeated letters in a password?

a. Does the order of the letters matter in the password? ____________

b. How many letters are there from *A* through *H*? ____________

c. Find the number of permutations of 8 letters taken 5 at a time.

$$_8P_5 = \frac{8!}{(8-5)!} = \_\_\_\_\_\_\_\_\_\_$$

8. An editor has 4 different spaces to arrange articles in a magazine. He must choose from 6 articles. How many different arrangements are possible?

Write and evaluate the permutation rule to solve.

____

Name ______________________ Date ____________ Class ____________

LESSON 11-1

Reteach

Permutations and Combinations (continued)

A **combination** is a selection of items from a group in which the order is NOT important. In a combination, *AB* is the same as *BA*.

The number of combinations of *n* items taken *r* at a time is shown by the following formula.

$$_nC_r = \frac{n!}{(n-r)!r!}$$

The combination rule is a modification of the permutation rule.

How many ways can club members select a committee of 4 people from a group of 10 members?
The order of the committee members does not matter. Selecting Tom and Pat is the same as selecting Pat and Tom.

To find the number of combinations of 10 items taken 4 at a time, use $n = 10$ and $r = 4$ in the combination rule. Then evaluate.

$$_{10}C_4 = \frac{10!}{(10-4)!4!} = \frac{10!}{6!4!} = \frac{10 \cdot 9 \cdot 8 \cdot 7 \cdot \cancel{6 \cdot 5 \cdot 4 \cdot 3 \cdot 2 \cdot 1}}{(\cancel{6 \cdot 5 \cdot 4 \cdot 3 \cdot 2 \cdot 1})(4 \cdot 3 \cdot 2 \cdot 1)}$$

$$= \frac{10 \cdot \overset{3}{\cancel{9}} \cdot \cancel{8} \cdot 7}{\cancel{4} \cdot \cancel{3} \cdot \cancel{2} \cdot \cancel{1}} = 10 \cdot 3 \cdot 7 = 210$$

Divide out common factors to simplify.

There are 210 ways to select the committee members.

Evaluate.

9. $_5C_2$ ____________

10. $_7C_6$ ____________

11. $_9C_4$ ____________

Solve.

12. A pollster has the names of 8 people available to answer her questions. She must select 3 of the people to interview. How many ways can she select the respondents?

    a. Does the order of the respondents matter? ____________

    b. Find the number of combinations of 8 people taken 3 at a time.

    $$_8C_3 = \frac{8!}{(8-3)!3!} = \text{\_\_\_\_\_\_\_\_\_\_}$$

13. The track team has 6 runners in a race. The coach selects 2 runners to run in the first heat. In how many ways can the runners be selected? Write and evaluate the combination rule to solve.

____

Name ______________________ Date __________ Class __________

LESSON 11-1

Challenge

Permutations with Special Conditions

A *permutation* is an arrangement of objects in a specific order. Sometimes there are also other conditions that must be satisfied. In such cases, you should deal with the special conditions first.

Using the letters in the word *square*, how many 6-letter arrangements with no repetitions are possible if vowels and consonants alternate, beginning with a vowel?

Of the 6 letters in the word, 3 are vowels (*u*, *a*, *e*) and 3 are consonants (*s*, *q*, *r*).

Beginning with a vowel, every other slot is to be filled by a vowel. There are 3 such slots and 3 vowels to be arranged in them.

3 × ___ × 2 × ___ × 1 × ___

The remaining 3 slots have 3 consonants to be arranged in them.

3 × 3 × 2 × 2 × 1 × 1

Multiply to determine the total number of arrangements.

There are 36 possible arrangements.

The girls Amy, Ann, and Doris and the boys Al, Arnon, Bob, and Roy are in a nursery group. Determine the number of ways the children can be arranged in a line with the following conditions.

1. A girl is always at the head of the line.

____

2. Roy is always at the head of the line.

____

3. A child whose name begins with *A* is always at the head of the line.

____

4. A child whose name begins with *A* is always at the head and the rear of the line.

____

The diamond suit from a standard deck of 52 playing cards is removed from the deck, shuffled, and laid out in a row. Determine the number of possible arrangements.

5. The first card is the ace.

____

6. The first card is a face card.

____

Use the digits 0, 1, 2, 3, 4 without repetition. Determine the number of ways to form each arrangement.

7. 3-digit numerals whose values are at least 100 ________

8. 4-digit numerals whose values are at least 1000 and less than 4000 ________

Name ______________________ Date __________ Class __________

LESSON 11-1

Problem Solving

Permutations and Combinations

Rosalie is looking at locks. The label *combination lock* confuses her. She wonders about the number of possible permutations or combinations a lock can have.

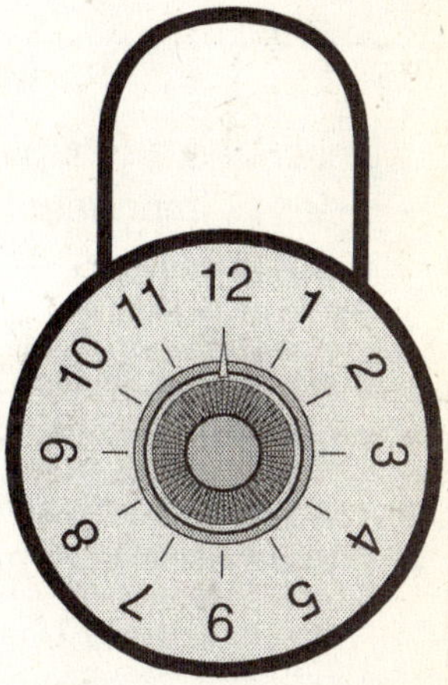

1. She looks at one circular lock with 12 positions. To open it she turns the dial clockwise to a first position, then counterclockwise to a second position, then clockwise to a third position

   a. Write an expression for the number of 3-position codes that are possible, if no position is repeated.

   ____

   b. Explain how this represents a combination or a permutation.

   ____

2. Rosalie looks at cable locks. Each position can be set from 0 to 9. How many different codes are possible for each lock if no digits are repeated in each code?

   a. a 3-digit cable lock

   ____

   b. a 4-digit cable lock

   ____

   c. a 6-digit cable lock

   ____

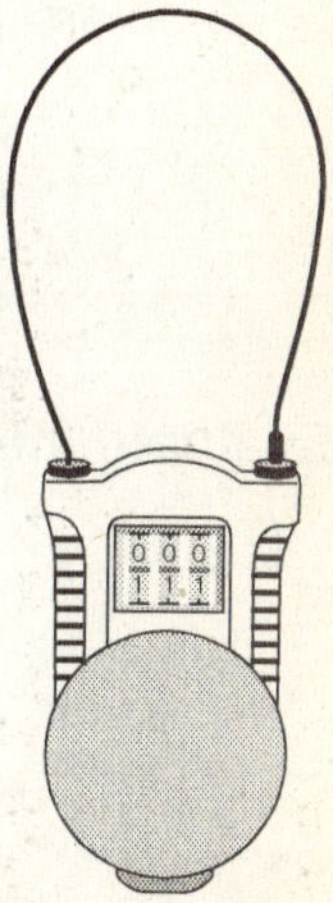

3. Rosalie needs 2 cable locks, but there are 13 types of locks to choose from.

   a. In how many ways can she choose 2 different locks? ______________

   b. Explain how this represents a permutation or a combination.

   ____

4. Explain why you think Rosalie might be confused by the label *combination lock.*

   ____

Rosalie wants to lock her bicycle near the library. There are 7 slots still open in the bike rack. Choose the letter for the best answer.

5. Rosalie arrives at the same time as 2 other cyclists. In how many ways can they arrange their bikes in the open slots?

   A 7 **C** 210

   B 35 **D** 343

6. Suppose Rosalie arrived just ahead of the 2 other cyclists and selected a slot. In how many ways can the others arrange their bikes in the open slots?

   F 2 **H** 24

   G 15 **J** 30

Name ______________________ Date ____________ Class ____________

LESSON 11-1

Reading Strategy
Compare and Contrast

Permutation	Combination
• Describes a grouping of objects in which order IS important	• Describes a grouping of objects in which order IS NOT important
• If John, Romeo, and Shelly all won prizes at the tennis tournament, it matters who won first prize, who won second prize, and who won third prize.	• If John, Romeo, and Shelly are chosen to compete in the tennis tournament, it doesn't matter in what order they are selected.
• The number of permutations of *x* items, chosen from a total of *n* items $_nP_x = \frac{n!}{(n-x)!}$	• The number of combinations of *x* items, chosen from a total of *n* items $_nC_x = \frac{n!}{x!(n-x)!}$
• *Example*: The number of ways that 3 objects can be selected from a total of 5 objects in a certain order. $_5P_3 = \frac{5!}{(5-3)!} = \frac{5 \cdot 4 \cdot 3 \cdot 2 \cdot 1}{2 \cdot 1} = 60$	• *Example*: The number of ways that 3 objects can be selected from a total of 5 objects in no particular order. $_5C_3 = \frac{5!}{3!(5-3)!} = \frac{5 \cdot 4 \cdot 3 \cdot 2 \cdot 1}{3 \cdot 2 \cdot 1(2 \cdot 1)} = 10$

Evaluate the following permutations and combinations.

1. $_4C_2$ ____________

2. $_4P_2$ ____________

3. $_5C_1$ ____________

4. $_5P_1$ ____________

5. $_3P_3$ ____________

6. $_7C_6$ ____________

Would you use a permutation or a combination to determine each of the following?

7. Nathan has a dozen eggs. He wants to decorate 4 eggs for an art project. ____________

8. Mr. Gunther is lining up 12 kindergarten students for a performance. ____________

9. Rachel has 10 valuable baseball cards. She wants to select 2 of them to sell online. ____________

10. Don has 5 soccer trophies to line up on the mantel of his fireplace. ____________

11. Bree has to select 5 photos from a box containing 25 photos to use in the yearbook. ____________

Name ______________________ Date ____________ Class ____________

LESSON 11-2

Practice A

Theoretical and Experimental Probability

Answer each question.

1. How many possible outcomes are there from tossing two number cubes labeled 1–6? ____________

2. Describe the sample space for a spinner with four equal sections of blue, red, green, and yellow. ____________

3. How likely is it that an outcome with a probability of 1 will occur? ____________

4. How likely is it that an outcome with a probability of 0 will occur? ____________

Solve.

5. A farmer has four sheepdogs and three beagles. If he randomly chooses a dog to accompany him on a walk, what is the probability of him taking a walk with a sheepdog? ____________

6. Gordon spins a spinner with equal-sized sections numbered 1–6. In one spin, what is the likelihood that the spinner will stop on a 1 or a 5? ____________

7. Oak trees shade 30% of the Fitzgeralds' backyard. What is the probability that someone standing at a random point in the backyard will NOT be in the shade? ____________

8. Find the probability that a point chosen at random inside the larger square shown here will also fall inside the smaller square.

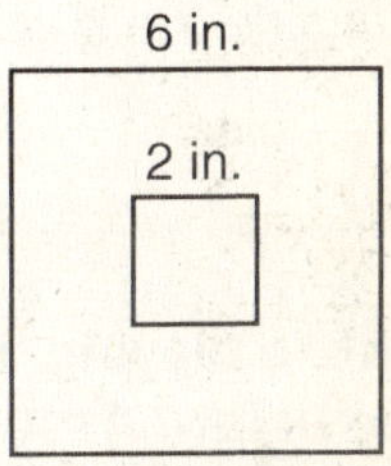

____

The table below shows the results of pulling one marble from a bag of marbles, recording its color, and replacing it in the bag.

Marble Color	Yellow	Red	Green
Times Pulled	53	17	30

Find the experimental probability of each event.

9. Choosing a yellow marble ____________

10. NOT choosing a red marble ____________

11. Choosing either a red or a green marble ____________

12. Which color marble is probably present in greatest number in the bag? ____________

Name ______________________ Date __________ Class __________

LESSON 11-2

Practice B

Theoretical and Experimental Probability

Solve.

1. A fruit bowl contains 4 green apples and 7 red apples. What is the probability that a randomly selected apple will be green? ____________

2. When two number cubes labeled 1–6 are rolled, what is the probability that the result will be two 4's? ____________

3. Joanne is guessing which day in November is Bess's birthday. Joanne knows that Bess's birthday does not fall on an odd-numbered day. What is the probability that Joanne will guess the correct day on her first try? ____________

4. Tom has a dollar's worth of dimes and a dollar's worth of nickels in his pocket.

   a. What is the probability he will randomly select a nickel from his pocket? ____________

   b. What is the probability he will randomly select a dime from his pocket? ____________

5. Clarice has 7 new CDs; 3 are classical music and the rest are pop music. If she randomly grabs 3 CDs to listen to in the car on her way to school, what is the probability that she will select only classical music? ____________

6. Find the probability that a point chosen at random inside the larger circle shown here will also fall inside the smaller circle.

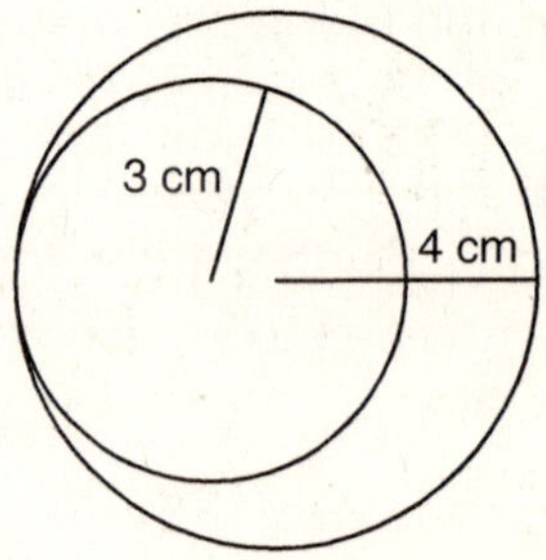

____

Frank is playing darts. The results of his throws are shown in the table below. Assume that his results continue to follow this trend.

Color Hit	Number of Throws
Blue	12
Red	5
White	2

Find the experimental probability of each event.

7. Frank's next throw will hit white. ____________

8. Frank's next throw will hit blue. ____________

9. Frank's next throw will hit either red or white. ____________

10. Frank's next throw will NOT hit red. ____________

Name ______________________ Date __________ Class __________

LESSON **11-2**

Practice C

Theoretical and Experimental Probability

Solve.

1. A bowl contains 36 blue, 75 green, and 19 yellow jelly beans. What is the probability of randomly selecting a green jelly bean? ______________

2. Two spinners numbered 1–6 are spun. If all numbers are equally likely, what is the probability that the result will be two even numbers? ______________

3. Four quilters are preparing patches for a quilt. When finished, the quilt will contain 200 patches. The quilters' contributions thus far are in the table below.

Name	Number of Patches
Lia	65
Brian	17
Elle	88
Len	6

   a. What is the probability that a randomly chosen patch will have been sewn by Elle? ______________

   b. What is the probability that a randomly chosen patch will not have been sewn by Lia? ______________

   c. What is the probability that a randomly chosen patch will have been sewn by Brian or Len? ______________

A hacker is trying to break into his school's computer system to change his F's to A's. The computer system access password is 5 digits.

4. If digits in the password are allowed to repeat, what is the probability that the hacker will guess the password correctly on the first try? ______________

5. The hacker learns that the password does not contain any repeated digits. What is the new probability that he will randomly guess the password correctly? ______________

6. If the password contains no repeated digits, what is the probability that the digits in the school password have a sum less than 10? ______________

Use the diagram to find each probability.

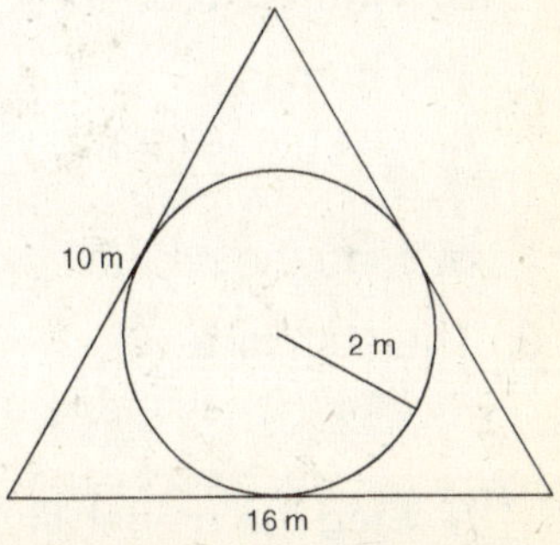

7. That a random point is within the circle in the triangle

____

8. That a random point is NOT within the circle in the triangle

____

Name ______________________ Date ____________ Class ____________

LESSON 11-2

Reteach

Theoretical and Experimental Probability

Probability measures the likelihood of an event, or **outcome**. All of the possible outcomes make up the **sample space**.

For outcomes that have the same likelihood of occurring, or equally likely outcomes, the probability of an event, *E*, is defined as follows.

$$P(E) = \frac{\text{number of favorable outcomes}}{\text{number of outcomes in the sample space}}$$

$0 \leq P(E) \leq 1$

Two 1–6 number cubes are rolled. If the numbers on each cube are equally likely, what is the probability that the sum is 8?

The greatest possible sum is 12.

Step 1 Find the sample space.
Use ordered pairs to list all the possible outcomes.
There are 36 outcomes in the sample space.

6	(6, 1)	**(6, 2)**	(6, 3)	(6, 4)	(6, 5)	(6, 6)
5	(5, 1)	(5, 2)	**(5, 3)**	(5, 4)	(5, 5)	(5, 6)
4	(4, 1)	(4, 2)	(4, 3)	**(4, 4)**	(4, 5)	(4, 6)
3	(3, 1)	(3, 2)	(3, 3)	(3, 4)	**(3, 5)**	(3, 6)
2	(2, 1)	(2, 2)	(2, 3)	(2, 4)	(2, 5)	**(2, 6)**
1	(1, 1)	(1, 2)	(1, 3)	(1, 4)	(1, 5)	(1, 6)
	1	2	3	4	5	6

Rows show outcomes of the second number cube.

Columns show outcomes of the first number cube.

Step 2 Find the outcomes with a sum of 8:
(6, 2), (5, 3), (4, 4), (3, 5), and (2, 6).
There are 5 outcomes with a sum of 8.

These are the favorable outcomes.

Step 3 Use the definition of theoretical probability.

$$P(\text{sum is 8}) = \frac{\text{number of outcomes with sum of 8}}{\text{total possible outcomes}} = \frac{5}{36}$$

Two number cubes are rolled. If all numbers are equally likely, what is the probability of each event?

1. The sum is 3.
   a. List the favorable outcomes. ______________________
   b. Find the probability. ______________________
2. The sum is 4.
   a. List the favorable outcomes. ______________________
   b. Find the probability. ______________________
3. The sum is 9.
   a. List the favorable outcomes. ______________________
   b. Find the probability. ______________________

Name ________________ Date ________ Class ________

LESSON 11-2

Reteach

Theoretical and Experimental Probability (continued)

To find the experimental probability of an event, conduct or observe an experiment for a large number of trials. Then count the number of times the event, *A*, occurs.

The **experimental probability** of the event is found by using the following formula.

$$P(A) = \frac{\text{number of times the event occurs}}{\text{number of trials}}$$

The table shows the results of an experiment with a spinner.

Spinner Experiment				
Color	Red	Yellow	Green	Blue
Spins	30	25	12	33

Outcomes (Color row)

Trials (Spins row)

Find the experimental probability of spinning red.

Step 1 First find the total number of trials.

$30 + 25 + 12 + 33 = 100$

Step 2 Use the definition of experimental probability.

You can use a fraction or a decimal.

$$P(\text{red}) = \frac{\text{number of times red occurred}}{\text{number of trials}} = \frac{30}{100} = \frac{3}{10} = 0.3$$

To find the probability of spinning green or blue, find the total number of times green or blue occurred.

$$P(\text{green or blue}) = \frac{\text{number of times green or blue occurred}}{\text{number of trials}}$$

$$= \frac{12 + 33}{100} = \frac{45}{100} = \frac{9}{20} = 0.45$$

The table shows the results of a spinner experiment. Use the table to find each experimental probability.

Spinner Experiment				
Number	1	2	3	4
Spins	15	10	8	22

4. Spinning a 3
   a. What is the total number of trials? ________________
   b. Find the probability. ________________
5. Spinning an even number
   a. What is the number of even spins? ________________
   b. Find the probability. ________________
6. Spinning 1 or 2
   a. How many trials resulted in a 1 or a 2? ________________
   b. Find the probability. ________________

Name ______________________ Date __________ Class __________

LESSON 11-2

Challenge

Using Monte Carlo Methods to Determine Area

The first means of generating random numbers used equipment similar to that in gambling casinos, so methods of using random numbers to simulate events are called *Monte Carlo methods*, named after the casino in the principality of Monaco. These methods can be used to approximate area under a curve.

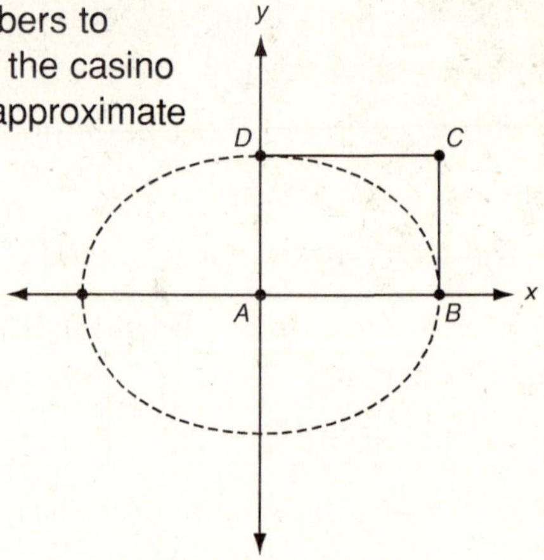

To approximate the area of the ellipse shown in the diagram, choose a point at random in rectangle *ABCD*. The probability that this point also lies in the first-quadrant region of the ellipse is the ratio of the area of the quarter of the ellipse to the area of rectangle *ABCD*. You can find this probability experimentally by generating *R* random points in rectangle *ABCD* and counting the number of points, *N*, that also lie in the ellipse.

As shown at right, set the area ratio approximately equal to $\frac{N}{R}$, the probability that a point lies inside the ellipse.

$$\frac{\text{area of quarter of ellipse}}{\text{area of rectangle } ABCD} \approx \frac{N}{R}$$

Use the equation $\frac{x^2}{9} + \frac{y^2}{4} = 1$ for the ellipse.

$$\text{area of quarter of ellipse} \approx 6\left(\frac{N}{R}\right)$$

Substitute the area, 6, for rectangle *ABCD* in the area ratio. Then multiply by 4 to determine the area of the entire ellipse.

$$\text{area of ellipse} \approx 24\left(\frac{N}{R}\right)$$

1. Use a graphing calculator to generate 25 points that lie in rectangle *ABCD*. Test whether these points lie inside the ellipse. Count the number of points, *N*, that are inside the ellipse.

a. Evaluate your ratio.

$\frac{N}{25}$ ______________________

b. Approximate the area of the ellipse.

____

c. Use the formula πab to calculate the area of the ellipse.

____

d. How can you refine your approximation?

____

Carry out a simulation to calculate the area. Repeat your simulation 2 times and take the average of the 3 results. Determine a theoretical value by using an area formula. Compare.

2. the area enclosed by the circle $x^2 + y^2 = 16$

____

3. the area enclosed by the ellipse $\frac{y^2}{64} + \frac{x^2}{25} = 1$ above the *x*-axis

____

Name ______________________ Date ____________ Class ____________

LESSON 11-2

Problem Solving

Theoretical and Experimental Probability

As part of a grant to improve bus routes to and from school, Hogan and Jane gather traffic flow statistics for one intersection. They make a table to show their findings for between 7:45 A.M. and 8:00 A.M. on a Monday morning.

Traffic Direction	Number of Vehicles
Straight through	282
Turn north	94
Turn south	188

1. Analyze the statistics.

   a. Write and evaluate an expression for $P(N)$, the probability that a vehicle will turn north.

   ____

   b. Write and evaluate an expression for the probability that a vehicle will turn north or go straight through the intersection.

   ____

   c. Write and evaluate an expression for the probability that a vehicle will not turn north.

   ____

2. The police department gathers statistics on Tuesday. Officers count a total of 608 vehicles, of which 380 go straight through the intersection, 76 turn north, and the rest turn south.

   a. What is the probability that a vehicle will turn north? ____________

   b. What is the probability that a vehicle will turn north or go straight through the intersection? ____________

   c. What is the probability that a vehicle will not turn north? ____________

3. Does this represent theoretical or experimental probability? Explain.

____

Math Assessment Survey				
Activity	Group projects	Keep a Journal	Multiple Choice	Word Problems
Student Response	57	18	35	10

A teacher surveys students on how they would prefer to have work assessed in math class. Choose the letter for the best answer.

4. What is the probability that a randomly chosen student prefers assessment through a group project?

   A $\frac{1}{12}$ C $\frac{19}{40}$

   B $\frac{53}{120}$ D $\frac{21}{40}$

5. Which expression gives the probability that a randomly chosen student will not want multiple-choice questions?

   F $\frac{7}{24}$ H $1 - \frac{7}{24}$

   G $1 + \frac{7}{24}$ J $35 - \frac{7}{24}$

Name ______________________ Date __________ Class __________

LESSON 11-2

Reading Strategy

Understand Vocabulary

You can predict the likelihood that an event will take place by calculating the **probability** (P) of it occurring. In order to calculate the probability of an event occurring, you need to know all the possible **outcomes** of an event.

$$P = \frac{n}{t}$$

n: The number of favorable outcomes

t: The number of possible outcomes

If you toss a dime, there are 2 possible outcomes: heads and tails.

If you roll a number cube, there are 6 possible outcomes: 1, 2, 3, 4, 5, and 6.

If you draw a card from a full deck, there are 52 possible outcomes.

If you randomly choose one day of the week, there are 7 possible outcomes.

Identify *n* and *t* in each statement.

1. There are 3 black kittens in a litter of 6 kittens. ____________
2. In a box of 8 donuts, 2 are chocolate. ____________
3. Half the children in the 16-student class are girls. ____________
4. Harry put a silver dollar in 4 of the 12 grab bags. ____________

The **complement** of an event is the set of all unfavorable outcomes.

$$P(\text{not } E) = 1 - P(E)$$

$P(E) + P(\text{not } E) = 1$

Answer each question.

5. What is the complement to rolling a 3 on a number cube? ____________
6. Fred has a bag containing 12 black marbles, 10 red marbles, and 15 white marbles.
   a. Fred wants to know the probability of randomly selecting a red marble from the bag. What are the unfavorable outcomes? ____________
   b. Write the formula to find the complement of choosing a red marble. ____________

Name ______________________ Date ____________ Class ____________

LESSON **11-3**

Practice A

Independent and Dependent Events

Find each probability.

1. Hal is tossing a quarter.

   a. What is the probability he will toss heads? ____________

   b. What is the probability he will toss tails? ____________

   c. What is the probability he will toss heads and then tails? ____________

2. Hal tosses a quarter three times. What is the probability the result will be tails each time? ____________

3. Katie rolls a 1–6 number cube twice. What is the probability she will roll an odd number and then an even number? ____________

4. Katie rolls the 1–6 number cube three times. What is the probability that the result will be a 3 each time? ____________

There are 3 apples and 5 oranges in a bag. Determine each probability.

5. Selecting 2 apples when they are chosen at random without replacement ____________

6. Selecting an orange, then an apple when they are chosen at random without replacement ____________

A student must have a B average or better for all courses to qualify for any athletic team at Jefferson High School. The table below shows the distribution of students' grades in three sports at the school.

Sport	Students with an A Average	Students with a B Average
Field hockey	15	4
Basketball	7	13
Football	2	22

An athlete is randomly selected. Find each probability in decimal form.

7. The student is a field hockey player with a B average. ____________

8. The student has an A average and plays football. ____________

9. The student has a B average and does NOT play football. ____________

There are 4 green marbles and 3 white marbles in a bag. A white marble is randomly selected and not replaced. Then a green marble is randomly selected.

10. Are these events dependent or independent? ____________

11. What is the probability of this event occurring? ____________

Name ______________________ Date ____________ Class ____________

LESSON 11-3

Practice B

Independent and Dependent Events

Find each probability.

1. A bag contains 5 red, 3 green, 4 blue, and 8 yellow marbles. Find the probability of randomly selecting a green marble, and then a yellow marble if the first marble is replaced. ____________

2. A sock drawer contains 5 rolled-up pairs of each color of socks, white, green, and blue. What is the probability of randomly selecting a pair of blue socks, replacing it, and then randomly selecting a pair of white socks? ____________

Two 1–6 number cubes are rolled—one is black and one is white.

3. The sum of the rolls is greater than or equal to 6 and the black cube shows a 3.

   a. Explain why the events are dependent.

   ____

   b. Find the probability. ____________

4. The white cube shows an even number, and the sum is 8.

   a. Explain why the events are dependent.

   ____

   b. Find the probability. ____________

The table below shows numbers of registered voters by age in the United States in 2004 based on the census. Find each probability in decimal form.

Age	Registered Voters (in thousands)	Not Registered to Vote (in thousands)
18–24	14,334	13,474
25–44	49,371	32,763
45–64	51,659	19,355
65 and over	26,706	8,033

5. A randomly selected person is registered to vote, given that the person is between the ages of 18 and 24. ____________

6. A randomly selected person is between the ages of 45 and 64 and is not registered to vote. ____________

7. A randomly selected person is registered to vote and is at least 65 years old. ____________

A bag contains 12 blue cubes, 12 red cubes, and 20 green cubes. Determine whether the events are independent or dependent, and find each probability.

8. A green cube and then a blue cube are chosen at random with replacement. ____________

9. Two blue cubes are chosen at random without replacement. ____________

Name ______________________ Date __________ Class __________

LESSON 11-3

Practice C

Independent and Dependent Events

Find each probability.

1. In cooking class, students are randomly choosing 1 of 3 different recipes. Two students choose the same recipe. ____________

2. Steven rolled a 1–6 number cube four times. The result was 4 odd numbers. ____________

The spinner shown here is spun twice.

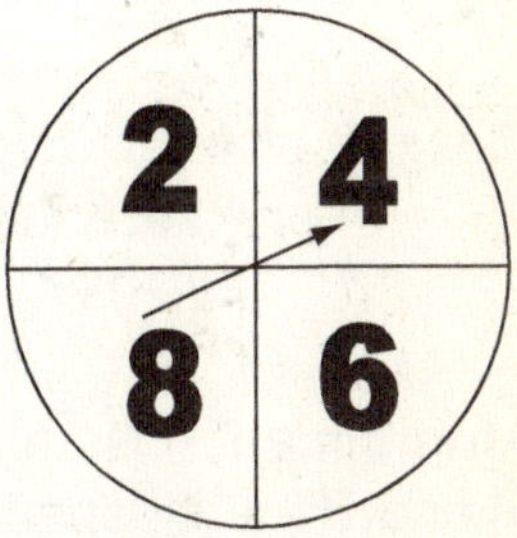

3. The sum of the results is equal to or greater than 10 and the first spin lands on 4.

   a. Explain why the events are dependent.

   ____

   b. Find the probability. ____________

4. The first spin lands on 6 and the sum of the results is less than or equal to 10.

   a. Explain why the events are dependent.

   ____

   b. Find the probability. ____________

The table shows the population distribution in Ireland in 1996 by age and gender.

Ireland's Population in 1996					
Age Group	0–20	21–40	41–60	61–80	Over 80
Males (in thousands)	620.4	526.8	405.3	212.0	33.0
Females (in thousands)	588.3	527.6	400.8	246.3	60.3

Find each probability.

5. A randomly selected person is no more than 20 years old, given that the person is male. ____________

6. A randomly selected person is female, given that the person is over 80 years old. ____________

A bag contains 3 red marbles, 7 yellow marbles, 5 green marbles, and 2 blue marbles. Determine whether the events are independent or dependent, and find each probability.

7. A red marble is chosen at random and replaced, and then the SAME red marble is chosen at random again. ____________

8. A yellow marble and then a blue marble are chosen at random without replacement. ____________

Name ______________________ Date __________ Class __________

LESSON 11-3

Reteach

Independent and Dependent Events

Two events, A and B, are **independent** if the occurrence of one does not affect the probability of the occurrence of the other.

Case 1: A card is drawn from a deck and then placed back in the deck. (Event A)
A second card is then drawn. (Event B)
Events A and B are independent.

Case 2: A card is drawn from a deck. It is not replaced. (Event C)
A second card is then drawn. (Event D)
Events C and D are NOT independent.

Multiplication Rule for the Probability of Independent Events

A and B are independent events.

$$P(A \text{ and } B) = P(A) \cdot P(B)$$

A deck of cards has 12 face cards and 40 number cards. A card is drawn from a deck and then placed back in the deck. A second card is then drawn. (The events are independent.)

What is the probability of drawing two face cards from the deck?

Step 1 Find the total number of cards. $12 + 40 = 52$

Step 2 Find the probability of drawing a face card.

$$P(\text{face card}) = \frac{12}{52} = \frac{3}{13}$$

Step 3 Use the rule for the probability of independent events.

$$P(2 \text{ face cards}) = P(\text{face card}) \cdot P(\text{face card})$$
$$= \frac{3}{13} \cdot \frac{3}{13} = \frac{9}{169}$$

What is the probability of drawing a face card and then a number card from the deck?

$$P(\text{number card}) = \frac{40}{52} = \frac{10}{13}$$

$$P(\text{face card, then number card}) = P(\text{face card}) \cdot P(\text{number card})$$
$$= \frac{3}{13} \cdot \frac{10}{13} = \frac{30}{169}$$

Find each probability.

1. Ben rolls a 4 and then a 5 on a 1–6 number cube.

a. $P(4)$ ______ **b.** $P(5)$ ______ **c.** $P(4, \text{then } 5) = P(4) \cdot P(5)$ ______

2. Ben rolls a 3 and then an even number on a 1–6 number cube.

a. $P(3)$ ______ **b.** $P(\text{even number})$ ______ **c.** $P(3) \cdot P(\text{even number})$ ______

Name ______________________ Date __________ Class __________

LESSON 11-3

Reteach

Independent and Dependent Events (continued)

Two events, *A* and *B*, are **dependent** if the occurrence of one affects the probability of the occurrence of the other.

A card is drawn from a deck. It is not replaced. — Event *A*
A second card is then drawn. — Event *B*
Events *A* and *B* are **dependent**.

The probability of *B* given that *A* has occurred is called the **conditional probability** of *B*, given *A*. It is denoted $P(B \mid A)$.

Multiplication Rule for the Probability of Dependent Events

A and *B* are dependent events.

$$P(A \text{ and } B) = P(A) \cdot P(B \mid A)$$

A deck of cards has 12 face cards and 40 number cards. A card is drawn from a deck and NOT replaced. A second card is then drawn. What is the probability of drawing two face cards from the deck? — The events are dependent.

Step 1 Find the probability of drawing a face card on the first draw. — Event *A*

$P(\text{face card}) = \frac{12}{52} = \frac{3}{13}$

Step 2 Find the probability of drawing a face card on the second draw. — Event *B*

$P(\text{face card} \mid \text{face card}) = \frac{11}{51}$ — If 1 face card was drawn, 11 face cards remain and 51 total cards remain.

Step 3 Use the rule for the probability of dependent events.

$P(\text{2 face cards}) = P(\text{face card}) \cdot P(\text{face card} \mid \text{face card})$

$= \frac{3}{13} \cdot \frac{11}{51} = \frac{33}{663} = \frac{11}{221}$

What is the probability of drawing a face card and then a number card from the deck?

$P(\text{number card} \mid \text{face card}) = \frac{40}{51}$ — There are still 40 number cards in the deck, but only 51 total cards remain.

$P(\text{face card, then number card}) = P(\text{face card}) \cdot P(\text{number card} \mid \text{face card})$

$= \frac{3}{13} \cdot \frac{40}{51} = \frac{40}{221}$

Find each probability.

3. Draw two number cards without replacing the first card.
   a. $P(\text{number card})$ ______________
   b. $P(\text{number card} \mid \text{number card})$ ______________
   c. $P(\text{number card}) \cdot P(\text{number card} \mid \text{number card})$ ______________
4. Draw a number card and then a face card without replacing the first card.
   a. $P(\text{number card})$ ______________
   b. $P(\text{face card} \mid \text{number card})$ ______________
   c. $P(\text{number card}) \cdot P(\text{face card} \mid \text{number card})$ ______________

Name ______________________ Date ____________ Class ____________

LESSON **11-3**

Challenge

Random Walks

The diagram at right shows a method of graphically recording the results of 6 coin tosses that occur in one-second intervals. The horizontal axis shows time, *t*, and the vertical axis shows position, *s*. Begin at the origin and move one unit up to record "heads" and one unit down to record "tails."

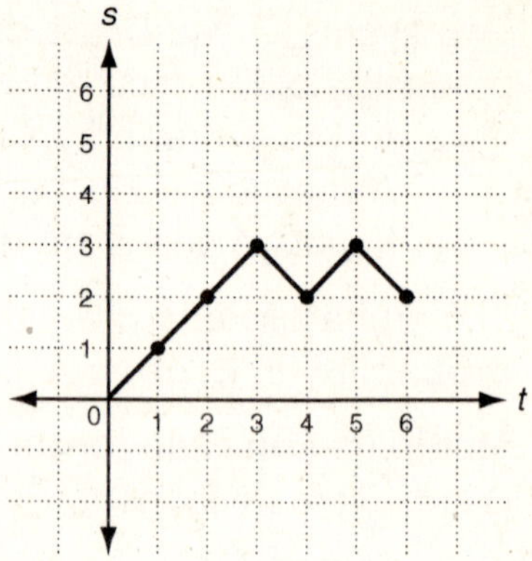

This type of graph is called a *random walk*.

1. Using *H* for "heads" and *T* for "tails," write the sequence for the random walk shown in the diagram at right.

____

Graph the random walk for the given coin sequences *A*, *B*, and *C*.

2. *A*: HHHHTTTHTHTTTTH

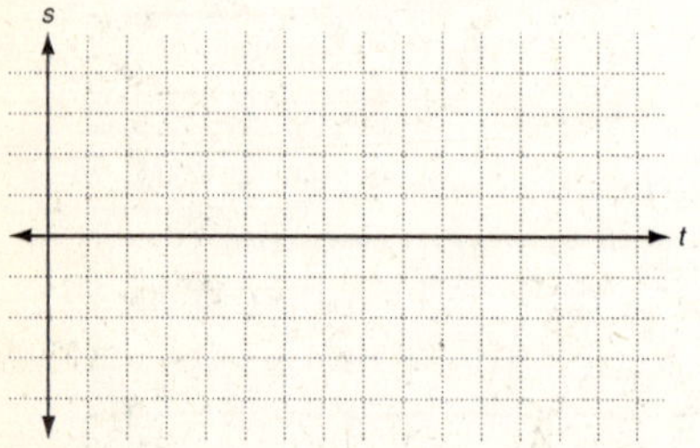

3. *B*: TTHTTHHTHTTHHHH

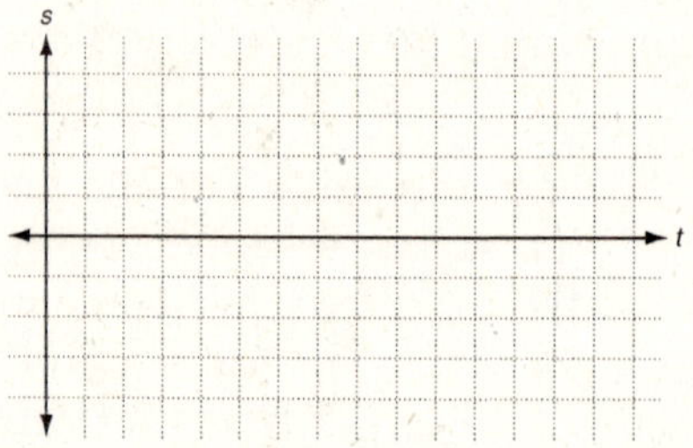

4. *C*: HHHHTTHTTHHHTTT

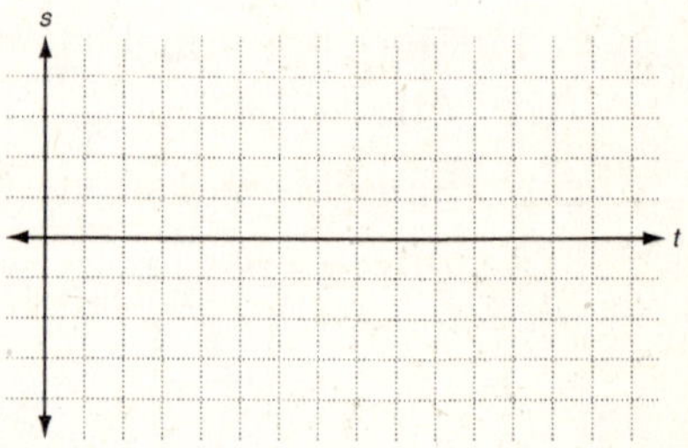

Refer to the graphs of sequences *A*, *B*, and *C*. For each sequence, give the time, if it exists, at which the random walk first returns to position $s = 0$. Then give the amount of time that the random walk spends in the first quadrant.

5. Sequence *A*

____

6. Sequence *B*

____

7. Sequence *C*

____

Solve.

8. With a partner, toss a coin 20 times to generate a random walk. Generate 5 such walks.

a. In what percent of your walks do you return to position $s = 0$ during the walk?

____

b. What is the average number of tosses to return to position $s = 0$?

____

9. Give an example of a real-life situation for which a random walk would be an appropriate model.

____

Name ______________________ Date __________ Class __________

LESSON 11-3

Problem Solving

Independent and Dependent Events

The table shows student participation in different sports at a high school. Suppose a student is selected at random.

Sports Participation by Grade					
	Track	Volleyball	Basketball	Tennis	No Sport
Grade 9	12	18	15	9	66
Grade 10	6	20	12	2	95
Grade 11	15	11	8	5	61
Grade 12	7	6	10	12	50

1. What is the probability that a student is in grade 10 and runs track?

   a. Find the probability that a student is in grade 10, $P(10)$. __________

   b. Find the probability that a student runs track, given that the student is in grade 10, $P(Tr \mid 10)$, __________

   c. Find $P(10 \text{ and } Tr) = P(10) \cdot P(Tr \mid 10)$. __________

2. What is the probability that a student is in grade 12 and runs track or plays tennis?

   a. Find the probability that a student is in grade 12, $P(12)$. __________

   b. Find the probability that a student runs track or plays tennis, given that the student is in grade 12, $P(Tr \text{ or } Te \mid 12)$. __________

   c. Find $P(12 \text{ or } (Tr \text{ or } Te))$. __________

3. During a fire drill, the students are waiting in the parking lot. What is the probability that one student is in grade 12 and runs track or plays tennis, and the student standing next to her is in grade 10 and runs track?

   a. Find the probability for the first student. __________

   b. Find the probability for the second student. __________

   c. Find the probability for the event occurring. __________

   d. Are these events independent or dependent? Explain. __________

____

Samantha is 1 of 17 students in a class of 85 who have decided to pursue a business degree. Each week, a student in the class is randomly selected to tutor younger students. Choose the letter for the best answer.

4. What is the probability of drawing a business student one week, replacing the name, and drawing the same name the next week?

   A 3.4 **C** 0.04

   B 0.2 **D** 0.002

5. What is the probability of drawing Samantha's name one week, not replacing her name, and drawing the name of another business student the next week?

   F $\frac{1}{85} \cdot \frac{16}{84}$ **H** $\frac{17}{85} \cdot \frac{16}{84}$

   G $\frac{1}{85} \cdot \frac{17}{84}$ **J** $\frac{17}{85} \cdot \frac{17}{84}$

Name ______________________________ Date ____________ Class ____________

LESSON 11-3

Reading Strategy

Draw Conclusions

In determining probability, events can be independent or dependent.

Events are independent if the occurrence of one does not affect the probability of the other.

Events are dependent if the occurrence of one affects the probability of the other.

A bag contains 4 apples and 4 oranges. Choosing a piece of fruit, replacing it, and choosing again are **independent** events. In each case there is the same number of each type of fruit in the bag.	Choosing a piece of fruit and not replacing it changes the probability of what you will choose the second time. There are now only 7 pieces of fruit in the bag. These are **dependent** events.

In both cases, calculate the probability of two events by multiplying the probability of each event occurring. However, the calculation of the probability of the second event requires a conclusion as to whether the events are independent or dependent.

Determine if the events are dependent or independent.

1. Tossing heads 3 times in a row ____________
2. Rolling a 5 on a number cube and then rolling another 5 ____________
3. Choosing a red candy from a bag of multicolored candies, not replacing it, and then choosing a green candy from the bag ____________
4. Spinning a spinner and having it land on an odd number, and spinning it again and having it land on an even number ____________
5. Drawing a red 8 from a deck of cards, replacing it, and then drawing a black queen ____________
6. Drawing a face card from a deck of cards, not replacing it, and then drawing an ace ____________
7. Choosing a senior from the track team to run in the state relay race, and then choosing a junior from the track team to run in the local 10K race ____________

Which is the more likely event?

8. Event 1: Choosing a green marble from a bag containing 3 green marbles and 2 blue marbles
   Event 2: Rolling a 4 on a number cube ____________
9. Event 1: Choosing the black crayon from a box of 20 crayons, replacing it, and choosing the white crayon
   Event 2: Choosing the black crayon from a box of 20 crayons, not replacing it, and choosing the white crayon ____________

Name ______________________ Date __________ Class __________

LESSON 11-4

Practice A
Compound Events

Determine which event(s) are mutually exclusive.

1. Students are forming 4 softball teams by each picking one of four different-color cards from a bag. Are the events "choosing a red card" and "choosing a blue card" mutually exclusive? Explain why or why not.

____

2. Are the events "choosing a black card" from a deck of playing cards and "choosing a 10" mutually exclusive? Explain why or why not.

____

Solve.

3. Apples are in $\frac{1}{4}$ of all of the lunch bags that are distributed at a school picnic, and bananas are in $\frac{1}{3}$ of the bags. What is the probability of randomly choosing a lunch bag that contains either an apple or a banana? __________

4. A group of senior citizens have won free vacation packages. The vacation to Bermuda is chosen by 25% of them, 60% choose Alaska, and 15% choose Costa Rica. What is the probability that one randomly selected senior citizen chooses to vacation in Costa Rica or Bermuda? __________

Pam rolls a 1–6 number cube. Find each probability.

5. Pam rolls a 3 or a 6. __________

6. Pam rolls an even number. __________

7. Pam rolls an odd or even number. __________

8. Pam rolls an odd number or a 3. __________

9. Pam rolls an odd number or a number greater than 4. __________

Use the data to fill in the Venn diagram. Then solve.

10. Of the 65 students going on the soccer trip, 43 are players and 12 are left-handed. Only 5 of the left-handed students are soccer players. What is the probability that one of the students on the trip is a soccer player or is left-handed? __________

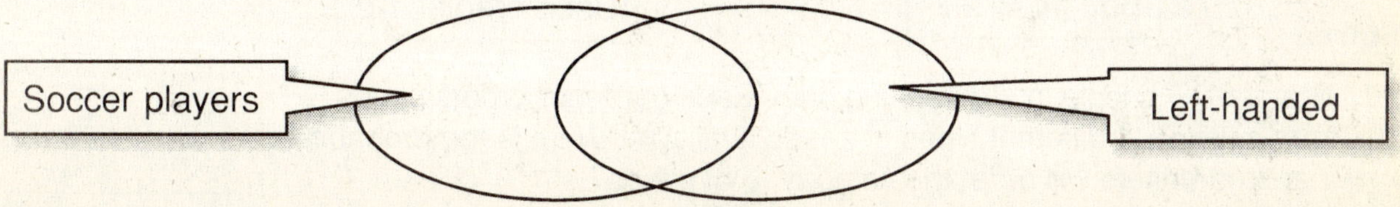

Name ______________________________ Date ___________ Class ___________

LESSON 11-4

Practice B

Compound Events

A can of vegetables with no label has a $\frac{1}{8}$ chance of being green beans and a $\frac{1}{5}$ chance of being corn.

1. Explain why the events "green beans" or "corn" are mutually exclusive.

____

2. What is the probability that an unlabeled can of vegetables is either green beans or corn? ___________

Ben rolls a 1–6 number cube. Find each probability.

3. Ben rolls a 3 or a 4. ___________

4. Ben rolls a number greater than 2 or an even number. ___________

5. Ben rolls a prime number or an odd number. ___________

Of the 400 doctors who attended a conference, 240 practiced family medicine and 130 were from countries outside the United States. One-third of the family medicine practitioners were not from the United States.

6. What is the probability that a doctor practices family medicine or is from the United States? ___________

7. What is the probability that a doctor practices family medicine or is not from the United States? ___________

8. What is the probability that a doctor does not practice family medicine or is from the United States? ___________

Use the data to fill in the Venn diagram. Then solve.

9. Of the 220 people who came into the Italian deli on Friday, 104 bought pizza and 82 used a credit card. Half of the people who bought pizza used a credit card. What is the probability that a customer bought pizza or used a credit card?

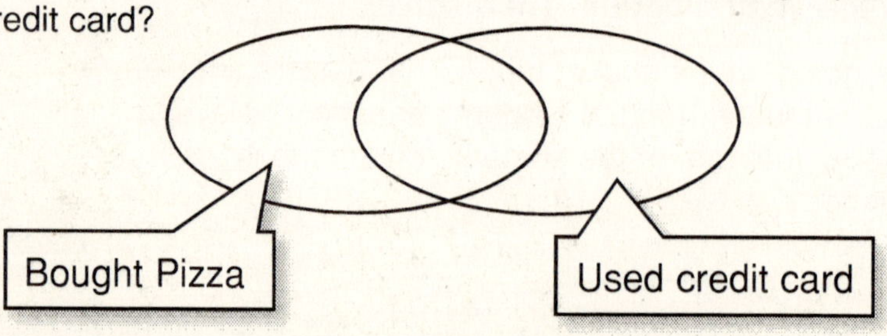

____

Solve.

10. There are 6 people in a gardening club. Each gardener orders seeds from a list of 11 different types of seeds available. What is the probability that 2 gardeners will order the same type of seeds? ___________

Name ______________________ Date ____________ Class ____________

LESSON **11-4**

Practice C
Compound Events

Cards numbered 1–25 are placed in a bag and one is drawn at random. Find each probability.

1. Drawing an odd number or a multiple of 7. ____________

2. Drawing an even number or a perfect square. ____________

A drug company is testing the side effects of different doses of a new drug on three different groups of volunteers.

Group	Volunteers	Daily Amount (mg)
A	353	150
B	467	200
C	310	250

3. If a volunteer is chosen randomly, what is the probability that this person receives the highest amount per day? ____________

4. If a volunteer is chosen randomly, what is the probability that this person receives more than 150 mg per day? ____________

5. If a volunteer is chosen randomly, what is the probability that this person does NOT receive 200 mg per day? ____________

Mr. Rodney's English class is made up of 28 students. He has 6 ESL students, 10 remedial students, and 5 advanced learners. ESL students make up $\frac{1}{5}$ of the remedial students and $\frac{3}{5}$ of the advanced learners.

6. What is the probability that a student is ESL and remedial? ____________

7. What is the probability that a student is ESL and an advanced learner? ____________

8. What is the probability that a student is remedial and NOT ESL? ____________

Solve.

9. A student is collecting a population of laboratory mice to be used in an experiment. He finds that of the 236 mice in the lab, 173 mice are female and 99 have pink eyes. Just 10 of the pink-eyed mice are male. What is the probability that a mouse is female or has pink eyes?

____

10. A group of 4 friends buys a CD of computer screen savers. If there are a total of 12 screen savers on the CD, what is the probability that at least 2 of the friends will choose the same screen saver for their computer?

____

Name ______________________________ Date __________ Class __________

LESSON 11-4

Reteach

Compound Events

A **simple event** is an event that cannot be broken down any further.
A **compound event** is an event made up of two or more simple events.

Rolling a 6 on a number cube is a simple event, while rolling an even number is a compound event.

Mutually exclusive events are events that cannot occur at the same time.

Tossing heads and tossing tails are mutually exclusive events. They cannot happen on the same toss of a coin.

Addition Rule for the Probability of Mutually Exclusive Events
A and *B* are mutually exclusive events.

$$P(A \text{ or } B) = P(A) + P(B)$$

A card is drawn from the deck. What is the probability of drawing a 5 or a face card from the deck?

Step 1 Decide whether the events are mutually exclusive.
A card cannot be both 5 and a face card.
The events are mutually exclusive.

Step 2 Find the probability of drawing a 5.
$P(5) = \frac{4}{52} = \frac{1}{13}$

There are four 5's in a deck of 52 cards.

Step 3 Find the probability of drawing a face card.
$P(\text{face card}) = \frac{12}{52} = \frac{3}{13}$

There are 12 face cards in a deck of 52 cards.

Step 4 Use the rule for the probability of mutually exclusive events.
$P(5 \text{ or face card}) = P(5) + P(\text{face card})$
$= \frac{1}{13} + \frac{3}{13} = \frac{4}{13}$

Toni draws a card from a deck. Find each probability.

1. She draws a number card or a king.

a. $P(\text{number card})$ ______________

b. $P(\text{king})$ ______________

c. $P(\text{number card or king}) = P(\text{number card}) + P(\text{king})$ ______________

2. She draws an ace or a queen.

a. $P(\text{ace})$ ______________

b. $P(\text{queen})$ ______________

c. $P(\text{ace or queen}) = P(\text{ace}) + P(\text{queen})$ ______________

Name ______________________ Date __________ Class __________

LESSON 11-4

Reteach

Compound Events (continued)

Inclusive events are compound events that can occur at the same time.

Rolling a 6 and rolling an even number on a number cube are inclusive events because 6 is an even number.

Because inclusive events can happen simultaneously, you do not want to double count the probability when the events overlap.

Addition Rule for the Probability of Inclusive Events
A and *B* are inclusive events.

$$P(A \text{ or } B) = P(A) + P(B) - P(A \text{ and } B)$$

$P(A \text{ and } B)$ denotes the probability that both *A* and *B* occur at the same time.

A card is drawn from the deck. What is the probability of drawing a 3 or a heart from the deck?

Step 1 Decide whether the events are inclusive.
The events overlap: there is a card that is both a 3 and a heart.
The events are inclusive.

Step 2 Find the probability of drawing a 3.
$P(3) = \frac{4}{52}$

There are four 3's in a deck of 52 cards.

Step 3 Find the probability of drawing a heart.
$P(\text{heart}) = \frac{13}{52}$

There are 13 hearts in a deck of 52 cards.

Step 4 Find the probability of drawing a 3 and a heart.
$P(3 \text{ and heart}) = \frac{1}{52}$

There is 1 card that is both a 3 and a heart.

Step 5 Use the rule for the probability of inclusive events.
$P(3 \text{ or heart}) = P(3) + P(\text{heart}) - P(3 \text{ and heart})$
$= \frac{4}{52} + \frac{13}{52} - \frac{1}{52} = \frac{16}{52} = \frac{4}{13}$

Charles draws a card from a deck. Find each probability.

3. He draws a red card or a king.

a. $P(\text{red})$ ______ **b.** $P(\text{king})$ ______ **c.** $P(\text{red and king})$ ______

d. $P(\text{red or king}) = P(\text{red}) + P(\text{king}) - P(\text{red and king})$ ______

4. He draws a diamond or a face card.

a. $P(\text{diamond})$ ______ **b.** $P(\text{face card})$ ______ **c.** $P(\text{diamond and face card})$ ______

d. $P(\text{diamond or face card}) = P(\text{diamond}) + P(\text{face card}) - P(\text{diamond and face card})$

______

Name ______________________________ Date ____________ Class ____________

LESSON 11-4

Challenge

Exploring P(A or B)

If A and B represent two events in the same sample space, then
$P(A \text{ or } B) = P(A) + P(B) - P(A \text{ and } B)$ when A and B are inclusive events.
When A and B are mutually exclusive events, this simplifies to $P(A \text{ or } B) = P(A) + P(B)$.
This formula can also be extended to apply to probabilities involving three events.

Consider the following situation.

A certain drug causes a skin rash or hair loss in 35% of patients. Twenty-five percent of patients experience only a skin rash, and 5% experience both a skin rash and hair loss. Find the probability that a patient will experience hair loss only.

1. Using A to represent "experiences a skin rash" and B to represent "experiences hair loss," write a symbolic representation of the problem.

   ____

2. Using the symbolic representation you wrote, write the answer to the question. ____________

Consider the following situation.

Tawana and Sergio recorded a compact disc together. Each sang some solos, and the two sang some duets. Sergio recorded twice as many duets as solos, and Tawana recorded six more solos than duets. When a CD player selects one of these songs at random, the probability that it will select a duet is 25%.

3. Let s represent the number of solos that Sergio recorded.
   a. Write an equation to express the probability of randomly selecting a duet in terms of s. ____________
   b. Solve the equation. Then determine the total number of songs recorded. ____________
   c. Find the probability of selecting a solo by Tawana or a duet when a CD player selects one song at random. ____________

4. Explain how the shaded area in the diagram shown below is used in the formula for $P(A \text{ or } B)$.

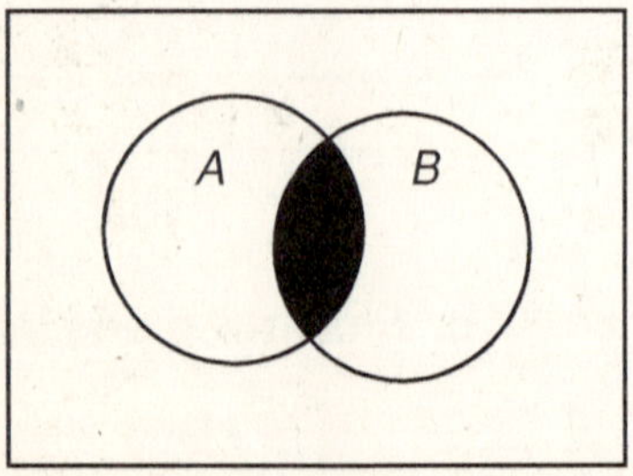

____

5. The diagram below represents three events in the same sample space. Use this diagram to write a formula for $P(A \text{ or } B \text{ or } C)$.

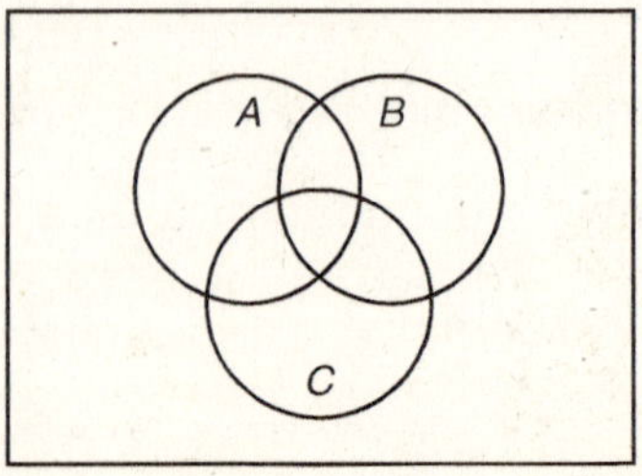

____

Name ________________________ Date ____________ Class ____________

LESSON **11-4**

Problem Solving

Compound Events

Of 100 students surveyed, 44 are male and 54 are in favor of a change to a 9-period, 4-day school week. Of those in favor, 20 are female. One student is picked at random from those surveyed.

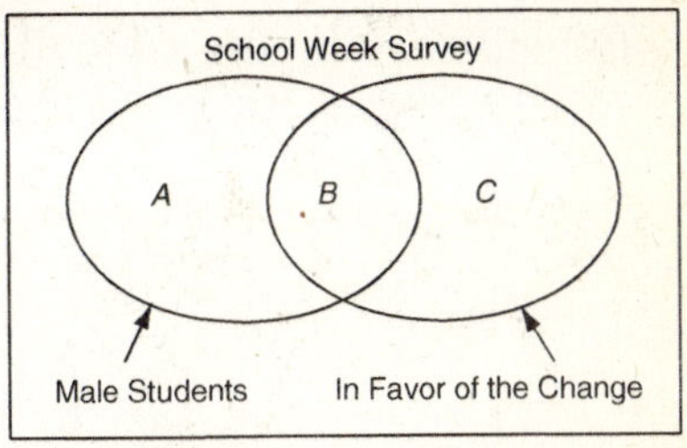

1. What is the probability that the student is male or favors the change? Use the Venn diagram.

   a. What is represented by the total of $A + B$?

   ____

   b. What is represented by the total of $B + C$?

   ____

   c. How many of those in favor of the change are male? ____________

   d. Find the values for A, B, and C and label the diagram.

   e. Write and evaluate an expression for the probability that the student is male or favors the change. ____________

2. What is the probability that the student is female or opposes the change?

   a. How many students are female? ____________

   b. How many students oppose the change? ____________

   c. If you draw a Venn diagram to show females and those opposed to the change, what is the meaning and value of the overlapping area?

   ____

   d. Write and evaluate an expression for the probability that the student is female or opposes the change. ____________

3. Of the students surveyed, 27 plan to start their own businesses. Of those, 18 are in favor of the change to the school week. Write and evaluate an expression for the probability that a student selected at random plans to start his or her own business or favors the change.

   ____

Sean asks each student to cast a vote for the type of class he or she would prefer. Of the students, 55% voted for online classes, 30% voted for projects, and 15% voted for following the textbook. Choose the letter for the best answer.

4. Which description best describes Sean's experiment?

   A Simple events

   B Compound events

   C Mutually exclusive events

   D Inclusive events

5. What is the probability that a randomly selected student voted for online classes or projects?

   F $\frac{33}{200}$ **H** $\frac{1}{4}$

   G $\frac{7}{10}$ **J** $\frac{17}{20}$

Name ______________________ Date ____________ Class ____________

LESSON **11-4**

Reading Strategy
Use a Venn Diagram

A Venn diagram can be used to organize what you know about mutually exclusive and inclusive events. The diagram on the left shows mutually exclusive events. There is no overlap so they can never occur at the same time. In the diagram on the right, the shaded overlapping region represents where both events A and B occur.

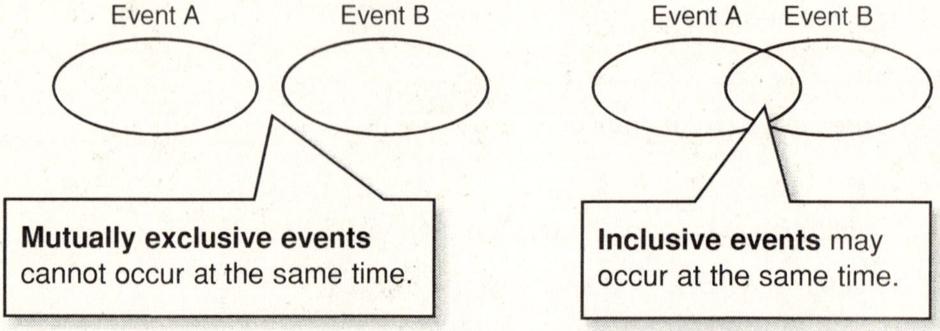

What is the probability of drawing a red card or a face card from a full deck?

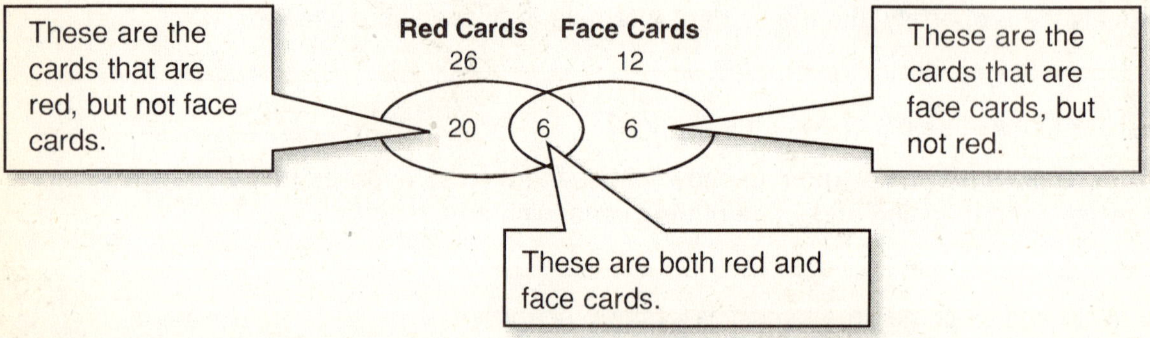

These are inclusive events because a card can be both red and a face card. Half of a deck of cards is red, so there are 26 red cards. There are 12 face cards in a deck. But some of the cards are both red and face cards. Of the 12 face cards, 6 of them are red. So, the overlapping region has a value of 6, since there are 6 cards that are favorable outcomes for both events A and B.

Draw and label a Venn diagram for each situation.

1. Of the 2250 voters surveyed, 1580 voted in the last election and 630 are planning to vote for Miller for mayor. Only 150 of those who voted in the last election are planning to vote for Miller.

2. A restaurant served 253 customers last night. The staff reported that 189 of the customers had made a reservation and that 163 customers used credit cards. Only 47 of those who used credit cards did not have a reservation.

Name ________________________ Date ____________ Class ____________

LESSON **11-5**

Practice A

Measures of Central Tendency and Variation

Find the mean, median, and mode of each data set.

1. {3, 10, 2, 8, 7, 5, 2, 5}

a. The mean is the sum of the values in a set divided by the number of values. Find the mean of the set. ____________

b. The median is the middle value when the set is ordered numerically. Find the median of the set. ____________

c. The mode is the value or values that occur most often. Find the mode of the set. ____________

2. {11, 15, 4, 10, 7, 5, 11, 9}

a. Mean ____________ **b.** Median ____________ **c.** Mode ____________

Find the expected values.

3. The probability distribution for the number of free throws that Larry makes in a game is given below. Find the expected number of free throws that Larry makes in a game. ____________

Number of Free Throws Made, *n*	5	6	7	8
Probability	0.21	0.58	0.14	0.07

4. The probability distribution for the number of pieces of junk mail May receives is given below. Find the expected number of junk mail letters May receives in a day. ____________

Number of Junk Mail Letters, *n*	1	2	3	4
Probability	0.15	0.70	0.10	0.05

Make a box-and-whisker plot of the data. Find the interquartile range.

5. {3, 7, 5, 3, 5, 9, 2, 7}

____

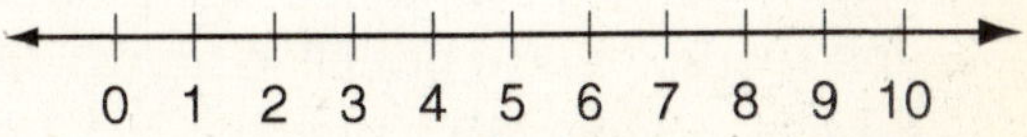

6. {1, 9, 9, 2, 2, 5, 5, 10}

____

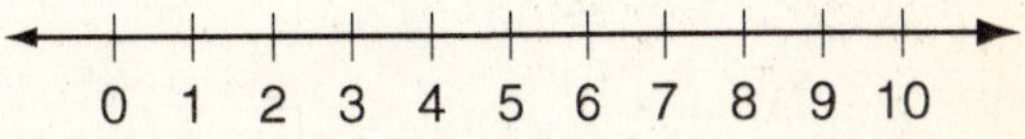

Find the variance and standard deviation.

7. {1, 2, 8, 11, 7, 10, 7, 2}

____

8. {10, 14, 8, 12, 9, 13}

____

Name ______________________ Date ____________ Class ____________

LESSON 11-5

Practice B

Measures of Central Tendency and Variation

Find the mean, median, and mode of each data set.

1. {12, 11, 17, 3, 9, 14, 16, 2}

a. Mean ____________________

b. Median ____________________

c. Mode ____________________

2. {6, 9, 9, 20, 4, 5, 9, 13, 10, 1}

a. Mean ____________________

b. Median ____________________

c. Mode ____________________

Make a box-and-whisker plot of the data. Find the interquartile range.

3. {3, 7, 7, 3, 10, 1, 6, 6}

____

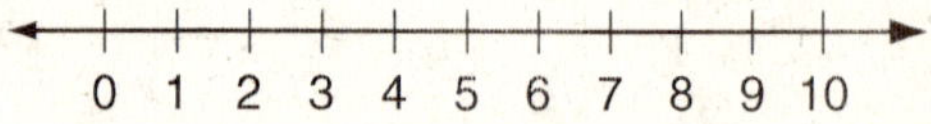

4. {1, 2, 3, 5, 3, 5, 8, 2}

____

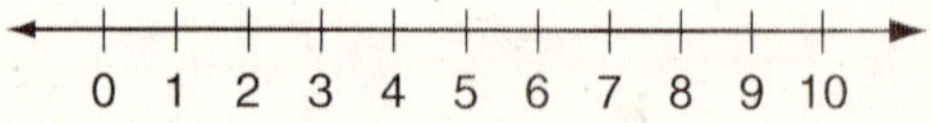

Find the variance and standard deviation.

5. {7, 4, 3, 9, 2}

____

6. {35, 67, 21, 16, 24, 51, 18, 32}

____

7. {19, 23, 17, 20, 25, 19, 15, 22}

____

8. {5, 12, 10, 13, 8, 11, 15, 12}

____

Solve.

9. The probability distribution for the amount of rain that falls on Boston in May each year is given below. Find the expected amount of rain for Boston in May. ____________________

Inches of Rain, *n*	5	6	7	8
Probability	0.05	0.10	0.64	0.21

10. A biologist is growing bacteria in the lab. For a certain species of bacteria, she records these doubling times: 41 min, 45 min, 39 min, 42 min, 38 min, 88 min, 43 min, 40 min, 44 min, 39 min, 42 min, and 40 min.

a. Find the mean of the data. ____________________

b. Find the standard deviation. ____________________

c. Identify any outliers. ____________________

d. Describe how any outlier affects the mean and the standard deviation.

____

Name ______________________ Date __________ Class __________

LESSON 11-5

Practice C

Measures of Central Tendency and Variation

Write a data set to satisfy the given conditions.

1. Median = 8; mode = 4

____

2. Mean = 10; median = 12

____

Make a box-and-whisker plot of the data. Find the interquartile range.

3. 21, 20, 4, 5, 5, 20, 20, 13, 1, 1, 13, 7, 13, 17, 7, 17, 9, 9 ______________

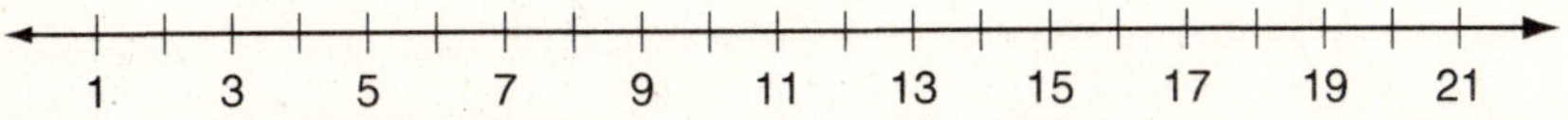

4. 10, 16, 3, 18, 18, 10, 8, 2, 9, 3, 7, 7, 8, 8, 18, 13, 7, 13 ______________

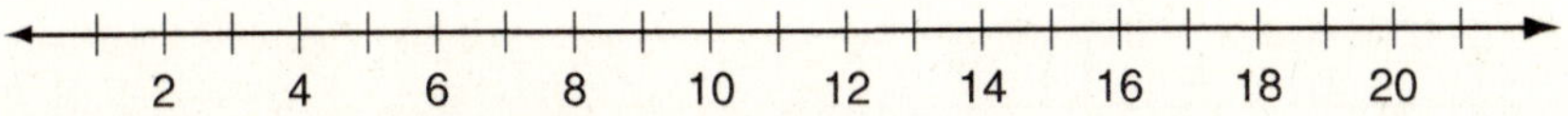

Find the variance and standard deviation.

5. {13, 7, 16, 22, 26, 11, 12, 19, 9}

____

6. {4, 7, 28, 6, 1, 1, 10, 15, 48, 3, 4, 5}

____

7. {15, 5, 12, 8, 19, 11, 7, 10, 9, 13, 17, 5}

____

8. {37, 29, 33, 30, 23, 28, 20, 35, 19, 21}

____

Solve.

9. The probability distribution for the number of children per family in a particular suburb of Chicago is shown below. Find the expected number of children per family in this region. ______________

Number of Children, *n*	1	2	3	4
Probability	0.27	0.28	0.35	0.10

10. A chemist weighs samples obtained from a production run. The weights of the samples are 13 g, 14 g, 65 g, 11 g, 15 g, 14 g, 14 g, 12 g, 13 g, 15 g, 14 g, and 12 g.

a. Find the mean of the data. ______________

b. Find the standard deviation. ______________

c. Identify any outliers. ______________

d. Describe how any outlier affects the mean and the standard deviation.

____

Name ______________________ Date __________ Class __________

LESSON 11-5

Reteach

Measures of Central Tendency and Variation

Expected value can be thought of as the long-term average of an experiment after many trials.

The table shows a probability distribution for the variable *x*. A probability distribution table pairs each outcome with its probability.

Outcome	x_1	x_2	x_3	...	x_n
Probability	p_1	p_2	p_3	...	p_n

Expected value $= x_1p_1 + x_2p_2 + x_3p_3 + \ldots + x_np_n$

To find the expected value, find the sum of the products of each outcome and its associated probability.

This table shows the probability distribution for the number of red marbles selected when 3 marbles are randomly selected without replacement from a bag with 4 red and 6 blue marbles. Find the expected number of red marbles selected.

Outcome	0	1	2	3
Probability	0.17	0.5	0.3	0.03

In the formula for the expected value, $n = 4$.

Expected value $= x_1p_1 + x_2p_2 + x_3p_3 + x_4p_4$

$= 0(0.17) + 1(0.5) + 2(0.3) + 3(0.03)$

$= 0 + 0.5 + 0.6 + 0.09 = 1.19$

The expected number of red marbles is 1.19.

Because the expected value is a type of average, it may not correspond exactly to any of the actual outcomes.

The probability distribution of temperatures, in degrees Celsius, from a chemistry experiment is shown below. Find the expected value of the temperature during the experiment.

Outcome	4	5	6	7	8
Probability	$\frac{2}{13}$	$\frac{4}{13}$	$\frac{3}{13}$	$\frac{4}{13}$	$\frac{2}{13}$

1. To use the formula for expected value, $n =$ ____________________.

2. Write the expected value formula to use.

____

3. Find the expected value of the temperature during the experiment.

____

Name ______________________ Date ___________ Class ___________

LESSON 11-5

Reteach

Measures of Central Tendency and Variation (continued)

The mean is a measure of central tendency. The variance and standard deviation are measures of dispersion or variation.
The notation often used for the mean is $\overline{x}$.
The variance is denoted σ^2 and the standard deviation is denoted σ.

This means they describe the spread of the data. The standard deviation measures spread from the mean.

Find the mean and standard deviation of the following quiz scores: 6, 8, 9, 8, 7, 5, 6, 7, 6, and 8.

Step 1 Find the mean.

$$\overline{x} = \frac{6+8+9+8+7+5+6+7+6+8}{10} = \frac{70}{10} = 7$$

Step 2 Find the difference between the mean and each data value.

Value of x	6	8	9	8	7	5	6	7	6	8
$x - \overline{x}$	−1	1	2	1	0	−2	−1	0	−1	1

Step 3 Square each value of $x - \overline{x}$.

$x - \overline{x}$	−1	1	2	1	0	−2	−1	0	−1	1
$(x - \overline{x})^2$	1	1	4	1	0	4	1	0	1	1

Step 4 Find the variance. Add all of the $(x - \overline{x})^2$ terms and divide by the number of data values.

$$\sigma^2 = \frac{1+1+4+1+0+4+1+0+1+1}{10} = \frac{14}{10} = 1.4$$

Step 5 Find the standard deviation.
The standard deviation is the square root of the variance.
$\sigma = \sqrt{1.4} \approx 1.18$

Dana's quiz scores are 2, 1, 5, 3, 1, 5, 3, and 4.

4. Find the mean. ____________

5. Find the difference between the mean and each data value: $x - \overline{x}$.

____

6. Find $(x - \overline{x})^2$ for each data value.

____

7. Find the variance.

____

8. Find the standard deviation.

____

Name ______________________ Date __________ Class __________

LESSON 11-5

Challenge

Survey Data Analysis

Cassie's job is to analyze a set of 100 integral value scores that range from 0 to 10 from a recent survey. The set of values includes exactly 6 scores that are zero. She decides not to include the 6 zero scores in the analysis, based on the false assumption that zero is not really a number since it has no value.

1. For the set of 94 values, excluding the zeros, Cassie correctly computes the mean. How does the mean for the 94 nonzero values compare to the mean for the 100 values? Is it greater than, less than, equal to, or is it impossible to determine? How do you know?

____

2. For the set of 94 values, excluding the zeros, Cassie correctly computes the median. How does the median for the 94 nonzero values compare to the median for the 100 values? Is it greater than, less than, equal to, or is it impossible to determine? How do you know?

____

3. For the set of 94 values, excluding the zeros, Cassie correctly computes the mode. How does the mode for the 94 nonzero values compare to the mode for the 100 values? Is it greater than, less than, equal to, or is it impossible to determine? How do you know?

____

4. For the set of 94 values, excluding the zeros, Cassie correctly computes the standard deviation. How does the standard deviation for the 94 nonzero values compare to the standard deviation for the 100 values? Is it greater than, less than, equal to, or impossible to determine? How do you know?

____

The frequency distribution for the 100 survey scores is shown below.

Survey Value	0	1	2	3	4	5	6	7	8	9	10
Frequency	6	8	9	11	13		12	10	8	6	5

5. How many people answered with the value of 5 in the survey? How do you know?

____

6. If you randomly chose 1 of the 100 people surveyed, what answer would you expect that person to give?

____

7. Find the first and third quartiles and the interquartile range for this set of values.

____

Name ______________________ Date ____________ Class ____________

LESSON **11-5**

Problem Solving

Measures of Central Tendency and Variation

Each week, Damien records the miles per gallon for his car, to the nearest whole number. Over a period of 10 weeks, the data are 18, 17, 19, 18, 18, 25, 29, 30, 26, 19. He wants to arrange and summarize his data so that he can analyze it.

1. Make a box-and-whisker plot of his data.

a. Order the data from least to greatest. ______________________

b. Identify the minimum, maximum, median, first quartile, and third quartile.

____

c. Use the number line to make a box-and-whisker plot of the data. Find and label the interquartile range.

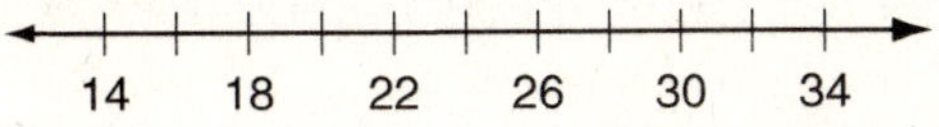

d. Explain what the interquartile range represents in terms of the car's miles per gallon.

____

2. Find the standard deviation for the data.

a. Write an equation and solve to find the mean. ______________________

b. Complete the table to show the difference between the mean and each data value, and the square of that difference.

Data Value, x	18	17	19	18	18	25	29	30	26	19
$x - \bar{x}$										
$(x - \bar{x})^2$										

c. Explain how to use the data from the table to find the standard deviation.

____

d. What is the standard deviation for the data? ______________________

e. Explain what the standard deviation represents in terms of the car's miles per gallon.

____

3. Damien thinks that the standard deviation is a more reliable measure of dispersion than the interquartile range. Is he correct? Explain.

____

____

____

Name ______________________ Date __________ Class __________

LESSON 11-5

Reading Strategy

Analyze Information

A box-and-whisker plot visually shows the distribution of data in a set. There are 9 members of the youth chess club. The following data set shows their ages.

{11, 17, 12, 16, 17, 14, 12, 9, 16}

The box-and-whisker plot of the data is shown below.

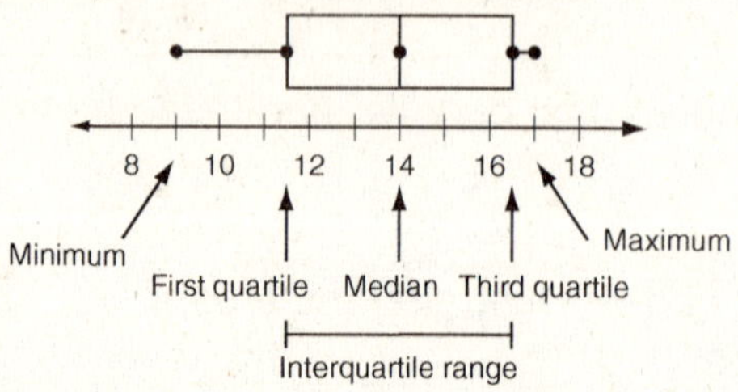

The median divides the data set into two parts, or quartiles. The median of the lower half of the data is the first quartile and the median of the upper half of the data is the third quartile. Where is the second quartile? It is the median of the full data set.

Answer each question.

1. The first quartile of some test scores is 55 and the interquartile range is 19.

a. Explain how you can find the third quartile.

____

b. Is it possible for the median of the test scores to be 77? Explain why or why not.

____

c. The maximum score is 93 and the range of the test scores is 55. What is the minimum test score?

____

d. Which quantity describes the middle 50% of the test scores? What is the range of test scores that will place a student in the middle 50% of the class?

____

2. Which data set is represented by the box-and-whisker plot shown?

A = {7, 9, 6, 12, 9, 15, 8, 14, 12, 10}

B = {6, 9, 7, 11, 8, 15, 11, 14, 10, 13}

C = {7, 8, 8, 12, 9, 13, 11, 14, 14, 15}

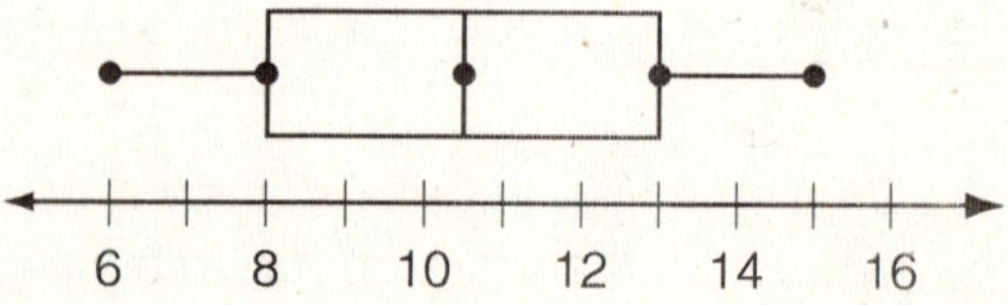

____

3. What do the "whiskers" in a box-and-whisker plot represent?

____

Name ______________________ Date ____________ Class ____________

LESSON 11-6

Practice A
Binomial Distributions

Use the Binomial Theorem to expand each binomial.

1. $(x + y)^2$

____

2. $(a + b)^3$

____

3. $(2x + y)^2$

____

4. $(x + 3y)^3$

____

Solve.

5. A coin lands heads half of the time it is tossed.

a. What is the probability that the coin will land heads? ____________

b. What is the probability that the coin will NOT land heads? ____________

c. If the coin is tossed 5 times, what is the probability that it will land heads twice?

____

d. The probability that the coin will land heads either once or twice is given by $P(1 \text{ or } 2) = P(1) + P(2)$. Find the probability.

____

6. A survey showed that 1 in 3 people entering the post office are wearing glasses. At noon, 4 people enter the post office.

a. What is the probability that one of those people is wearing glasses?

____

b. What is the probability that one of those people is NOT wearing glasses?

____

c. What is the probability that at least 2 of the 4 people are wearing glasses?

____

d. The probability that at least 3 of the people are wearing glasses is given by $P(\geq 3) = P(3) + P(4)$. Find the probability.

____

Name ______________________ Date ____________ Class ____________

LESSON 11-6

Practice B
Binomial Distributions

Use the Binomial Theorem to expand each binomial.

1. $(x + y)^3$

____

2. $(2x + y)^4$

____

3. $(m + 3n)^3$

____

4. $(p + q)^5$

____

Solve.

5. Of the new cars in a car dealer's lot, 1 in 6 are white. Today, 4 cars were sold.

a. What is the probability that 3 of the cars sold were white?

____

b. What is the probability that at least 2 of the cars sold were white?

____

6. At a small college, $\frac{1}{3}$ of all of the students are vegetarians. There are 5 students in line at the cafeteria.

a. What is the probability that all 5 students are vegetarians?

____

b. What is the probability that just 1 of the students is a vegetarian?

____

7. Ellen plays 8 hands of a card game with her friends. She has a 1 in 3 chance of winning each hand. What is the probability that she will win exactly half of the hands played?

____

8. In a lottery, each ticket buyer has a 1 in 10 chance of winning a prize. If Chip buys 10 tickets, what is the probability that he will win at least 1 prize?

____

Name ______________________ Date __________ Class __________

LESSON 11-6

Practice C

Binomial Distributions

Use the Binomial Theorem to expand each binomial.

1. $(x + y)^5$

____

2. $(4x + y)^4$

____

3. $(2x + y)^5$

____

4. $(n + 2m)^4$

____

Solve.

5. At Hopewell High School, 1 in 7 students is on a sports team. There are 4 student council representatives in the school.

a. What is the probability that 2 of the student council representatives are also on a sports team?

____

b. What is the probability that at least 3 of the student council representatives are on a sports team?

____

6. Two in every 5 donuts in a package of donuts have a jelly filling. There are 5 donuts left in the package.

a. What is the probability that all 5 donuts have a jelly filling?

____

b. What is the probability that none of the donuts has a jelly filling?

____

7. Andrew is choosing CDs from a bag of free CDs without looking. He has a 1 in 5 chance of choosing a CD that he likes. He chooses 8 CDs in all. What is the probability that he will get 3 CDs that he likes?

____

8. In a game of Bingo, the contestants have a 1 in 12 chance of winning each round. If Shirley plays 6 rounds, what is the probability that she will win at least half of them?

____

Name ______________________________ Date ____________ Class ____________

LESSON 11-6

Reteach

Binomial Distributions

The coefficients of the expansion of $(x + y)^n$ are the numbers in Pascal's triangle. The binomial coefficients are also combinations.

Pascal's Triangle	Combinations	Binomial Expansion
1	${}_0C_0$	$(x + y)^0 = 1$
1 1	${}_1C_0 \quad {}_1C_1$	$(x + y)^1 = x + y$
1 2 1	${}_2C_0 \quad {}_2C_1 \quad {}_2C_2$	$(x + y)^2 = x^2 + 2xy + y^2$
1 3 3 1	${}_3C_0 \quad {}_3C_1 \quad {}_3C_2 \quad {}_3C_3$	$(x + y)^3 = x^3 + 3x^2y + 3xy^2 + y^2$

The **Binomial Theorem** summarizes the relationship.

$$(x + y)^n = {}_nC_0x^ny^0 + {}_nC_1x^{n-1}y^1 + {}_nC_2x^{n-2}y^2 + \ldots + {}_nC_{n-1}x^1y^{n-1} + {}_nC_nx^0y^n$$

The sum of the exponents for each term is *n*.

You can use the Binomial Theorem to expand any binomial.

Expand $(x + y)^5$.

Step 1 Use the Binomial Theorem with n = 5.

$(x + y)^5 = {}_5C_0x^5y^0 + {}_5C_1x^4y^1 + {}_5C_2x^3y^2 + {}_5C_3x^2y^3 + {}_5C_4x^1y^4 + {}_5C_5x^0y^5$

Step 2 Use the fifth row of Pascal's triangle to find the coefficients. Continue the pattern and add pairs of terms from the fourth row to find the numbers in the fifth row.

The sum of the exponents for each term is 5. The exponents of *x* decrease as the exponents of *y* increase.

	1	3	3	1		← $n = 3$
1	4	6	4	1		← $n = 4$
1	5	10	10	5	1	← $n = 5$

Step 3 Write the coefficients.

$(x + y)^5 = 1x^5y^0 + 5x^4y^1 + 10x^3y^2 + 10x^2y^3 + 5x^1y^4 + 1x^0y^5$

Step 4 Simplify.

$(x + y)^5 = x^5 + 5x^4y + 10x^3y^2 + 10x^2y^3 + 5xy^4 + y^5$

Use the Binomial Theorem to expand each binomial.

1. $(2x + y)^3 = {}_3C_0(2x)^3y^0 + {}_3C_1(2x)^2y^1 + {}_3C_2(2x)^1y^2 + {}_3C_3(2x)^0y^3$

$= {}_3C_0 \cdot 8x^3y^0 + {}_3C_1 \cdot 4x^2y^1 + {}_3C_2 \cdot 2xy^2 + {}_3C_3 \cdot y^3$

= ____

= ____

2. $(x + 3y)^4$ = ____

= ____

= ____

Name ______________________ Date ____________ Class ____________

LESSON 11-6

Reteach

Binomial Distributions (continued)

A binomial experiment must meet the following conditions.

- It has a fixed number of trials, n.
- The trials are independent.
- The outcomes of the trials must fall into two categories.
- The probabilities remain constant for each trial.

The categories are classified as success and failure.

If a binomial experiment has n trials with probability of success p and probability of failure q for any trial, then the binomial probability of **exactly** r successes is given by the following formula.

$$P(r) = {}_nC_r p^r q^{n-r}$$

Remember that $q = 1 - p$.

In a game with a number cube, a person wins if the number 1 or 2 is rolled. The person loses if the number 3, 4, 5, or 6 is rolled. In 4 rolls, find the probability of exactly 2 wins.

The game satifies the conditions of a binomial experiment. There are 4 independent trials, each with fixed probability of success or failure.

Step 1 Find the probability of success (win) and failure (lose).

$$p = \frac{2}{6} = \frac{1}{3}$$

So $q = 1 - p = 1 - \frac{1}{3} = \frac{2}{3}$.

Step 2 Use the binomial probabilty formula with $r = 2$ successes, or wins, $n = 4$ rolls, or trials, $p = \frac{1}{3}$, and $q = \frac{2}{3}$.

$$P(2) = {}_4C_2\left(\frac{1}{3}\right)^2\left(\frac{2}{3}\right)^{4-2}$$

$$= {}_4C_2\left(\frac{1}{3}\right)^2\left(\frac{2}{3}\right)^2$$

$$= 6\left(\frac{1}{9}\right)\left(\frac{4}{9}\right)$$

$$= \frac{8}{27}$$

Think: ${}_4C_2$ is the second term in the fourth row of Pascal's triangle.

So, the probability is $\frac{8}{27}$ that 2 out of 4 rolls of a number cube will result either a 1 or a 2.

Diane and Ed are playing a game with a number cube. Diane wins if the number 6 is rolled. She loses if any other number is rolled. Find the probability of Diane having exactly 2 wins in 3 rolls.

3. Find p and q. ____________

4. Find n and r. ____________

5. Write the appropriate binomial probability formula. ____________

6. Find the probability. ____________

Name ______________________ Date ____________ Class ____________

LESSON 11-6

Challenge

The Dice Game

Antoine Gombauld, the Chevalier de Méré, was a seventeenth century French dice player. He was a successful gambler in one game of dice rolling and unsuccessful in another. He would bet that he could roll at least one 6 in 4 rolls of a single die. With this bet he made money in the long run. But he would also bet that he could roll at least one pair of 6s in 24 rolls of a pair of dice. He eventually lost money with this bet.

The first experiment is rolling at least one 6 in 4 rolls of a single die.

1. Is each of the 4 rolls of the single die independent of one another? Why or why not?

____

2. What is the probability of rolling at least one 6 in 4 rolls? This means rolling either one 6, two 6s, three 6s, or four 6s.

____

3. What is the probability of rolling no 6s in the 4 rolls? How does your answer compare to your previous answer?

____

4. Predict the outcome of Gombauld's bets with this game over time. How much might he win or lose for each $1 bet?

____

Consider the game of rolling at least one pair of 6s in 24 rolls of a pair of dice.

5. Is each of the 24 rolls of the pair of dice independent? Why or why not?

____

6. What is the probability of rolling at least one pair of 6s in 24 rolls of a pair of dice? How can you find this probability?

____

7. Predict the outcome of Gombauld's bets with this game over time. How much might he win or lose for each $1 bet?

____

____

8. How might Antoine Gombauld have changed the rules so that he could have had a positive expected return for this game?

____

____

Name ______________________________ Date __________ Class __________

LESSON **11-6**

Problem Solving

Binomial Distributions

Sales records for the snack machines show that 1 out of every 6 students buys a bag of trail mix. There are 5 students waiting to use the machines. Melanie uses the formula for binomial probability, $P(r) = {}_nC_r p^r q^{n-r}$, to determine the number of students expected to buy trail mix.

1. What is the probability of exactly 3 students buying a bag of trail mix?

   a. What is the probability of each student buying a bag of trail mix? ______________

   b. Define each variable used in the formula and give its value.

   ____

   c. Write the binomial formula using these values.

   d. Solve the equation to give the probability of exactly 3 students buying a bag of trail mix. ______________ ______________

2. What is the probability of at least 1 student buying a bag of trail mix?

   a. Describe a method to solve involving the sum of probabilities.

   ____

   b. Describe a method to solve that uses the formula $P(E) + P(\text{not } E) = 1$.

   ____

   c. Use either method to determine the probability of at least 1 student buying a bag of trail mix. ______________

3. After school, 4 students line up to buy snacks from the machine. What is the probability that they will all buy a bag of trail mix? ______________

Sports drinks are purchased by 3 out of 4 students using the snack machines. There are 3 students at the machines now. Choose the letter for the best answer.

4. Which expression gives the probability of exactly 2 students buying an energy drink?

   A $P(2) = {}_3C_2\left(\frac{3}{4}\right)^2\left(\frac{1}{4}\right)^1$

   B $P(2) = {}_4C_3\left(\frac{2}{3}\right)^2\left(\frac{1}{3}\right)^1$

   C $P(3) = {}_3C_2\left(\frac{3}{4}\right)^1\left(\frac{1}{4}\right)^2$

   D $P(3) = {}_4C_3\left(\frac{2}{3}\right)^1\left(\frac{1}{3}\right)^2$

5. What is the probability that at least 2 of the students will buy an energy drink?

   F 42%

   G 50%

   H 75%

   J 84%

Name ______________________ Date __________ Class __________

LESSON 11-6

Reading Strategy
Follow a Procedure

The probability that an event will happen r times can be determined by using a formula that involves the probability that the event will occur (p), the probability that it will not occur (q), a combination, and the total number of events taking place (n).

Example: In the senior class, 1 out of 4 students plays a sport. The principal chooses 2 students at random. What is the probability that both students play a sport?

Step 1	Identify n, p, q, and r.	$n = 2; p = \frac{1}{4}; q = 1 - \frac{1}{4} = \frac{3}{4}; r = 2$
Step 2	Write the equation and substitute for the variables.	Equation: $P(r) = {}_nC_r p^r q^{n-r}$ Substitute: $P(2) = {}_2C_2 \left(\frac{1}{4}\right)^2 \left(\frac{3}{4}\right)^0$
Step 3	Evaluate the combination.	${}_2C_2 = \frac{2!}{2!(2-2)!} = 1$
Step 4	Complete the solution.	$P(2) = 1\left(\frac{1}{4}\right)^2 \left(\frac{3}{4}\right)^0$ $= 1\left(\frac{1}{16}\right)1 = \frac{1}{16} \approx 0.0625$

Follow the procedure to solve.

1. Of the 6 kittens in the litter, 1 in 3 has blue eyes. What is the probability that if 2 kittens are picked at random, 1 will have blue eyes?

   a. Identify n, p, q, and r. ______________________

   b. Write the equation and substitute for the variables.

   ____

   c. Evaluate the combination. ______________________

   d. Complete the solution. ______________________

2. Jason sees that 2 out of every 5 cars that passed his house had windshield wipers on. What is the probability that of the next 4 cars that pass his house, 2 of them will have windshield wipers on?

   a. Identify n, p, q, and r. ______________________

   b. Write the equation and substitute for the variables.

   ____

   c. Evaluate the combination. ______________________

   d. Complete the solution. ______________________

LESSON 11-1

Practice A

Permutations and Combinations

Use the Fundamental Counting Principle.

1. For her aquarium, Susan can choose from 4 types of fish and 3 types of plants. If she chooses one type of fish and one type of plant, how many different aquariums can Susan set up? **12**

2. Lottery numbers in a particular state consist of 6 digits. Each lottery ball that is drawn has six sides, numbered 1–6. How many different lottery numbers are possible? **46,656**

Evaluate.

3. $5!$ **120**

4. $\frac{6!}{2!}$ **360**

5. $\frac{2!3!}{4!}$ $\mathbf{\frac{1}{2}}$

Solve.

6. For an art exhibit, Craig has to choose 3 ceramic mugs out of the 7 that he made over the summer. In how many ways can he arrange these 3 mugs in a row? **210 ways**

7. The members of a track team want to choose a team captain and someone to organize their equipment. In how many ways can 2 people be chosen from a team of 10 girls? **90 ways**

Evaluate.

8. ${}_4P_3$ **24**

9. ${}_3C_2$ **3**

10. ${}_5P_2$ **20**

11. ${}_5C_2$ **10**

12. ${}_8P_4$ **1680**

13. ${}_6C_1$ **6**

Solve.

14. While on vacation, Sandra wants to buy 2 wallets. There are 7 different patterns she can choose from. In how many ways can Sandra choose 2 different wallets? **21 ways**

15. Vince chooses 3 side dishes from a total of 10 side dishes offered on the menu. In how many different ways can he choose his side dishes? **120 ways**

LESSON 11-1

Practice B

Permutations and Combinations

Use the Fundamental Counting Principle.

1. The soccer team is silk-screening T-shirts. They have 4 different colors of T-shirts and 2 different colors of ink. How many different T-shirts can be made using one ink color on a T-shirt? **8 T-shirts**

2. A travel agent is offering a vacation package. Participants choose the type of tour, a meal plan, and a hotel class from the table below.

Tour	Meal	Hotel
Walking	Restaurant	4-Star
Boat	Picnic	3-Star
Bicycle		2-Star
		1-Star

How many different vacation packages are offered? **24 packages**

Evaluate.

3. $\frac{3!6!}{3!}$ **720**

4. $\frac{10!}{7!}$ **720**

5. $\frac{9! - 6!}{(9-6)!}$ **60,360**

Solve.

6. In how many ways can the debate team choose a president and a secretary if there are 10 people on the team? **90 ways**

7. A teacher is passing out first-, second-, and third-place prizes for the best student actor in a production of *Hamlet*. If there are 14 students in the class, in how many different ways can the awards be presented? **2184 ways**

Evaluate.

8. ${}_5P_4$ **120**

9. ${}_3C_2$ **3**

10. ${}_8P_3$ **336**

Solve.

11. Mrs. Marshall has 11 boys and 14 girls in her kindergarten class this year.

a. In how many ways can she select 2 girls to pass out a snack? **91 ways**

b. In how many ways can she select 5 boys to pass out new books? **462 ways**

c. In how many ways can she select 3 students to carry papers to the office? **2300 ways**

LESSON 11-1

Practice C

Permutations and Combinations

Evaluate.

1. $\frac{7! - 4!}{(6-3)!}$ **836**

2. $\frac{6!}{3!(8-5)!}$ **20**

3. $\frac{5!4!}{9!}$ $\mathbf{\frac{1}{126}}$

4. ${}_{10}C_5$ **252**

5. ${}_7P_4$ **840**

6. ${}_{10}C_9$ **10**

Compare. Write >, <, or =.

7. ${}_8P_3 \square {}_6C_3$ **>**

8. ${}_{12}C_9 \square {}_{12}C_7$ **<**

9. ${}_9P_5 \square {}_{10}P_5$ **<**

Solve.

10. The door code to get into a top-secret laboratory is 6 digits. The first 3 digits of the code are all odd and the last 3 digits are all even. Digits can be used more than once. How many possible codes are there to gain access to this laboratory? **15,625 codes**

11. In how many ways can a 3-digit number be formed using the numbers 0–9, if each digit is used only one time? **720 ways**

12. The principal of the high school selects 4 Merit Scholars to attend a Town Council meeting. If there are a total of 12 Merit Scholars at the school, in how many ways can the students be selected? **495 ways**

13. A board of trustees is made up of 10 people. The board is choosing a chairperson, a secretary, and a publicist. If they have already decided upon a chairperson, in how many ways can they choose a secretary and a publicist? **72 ways**

14. There are 8 marbles in a bag. If they are all different colors, in how many ways can 4 marbles be chosen? **70 ways**

15. A student in a biology laboratory has 7 plants to use in an experiment. One plant will act as the control, 3 will be subjected to Environment A, and 3 will be subjected to Environment B. In how many ways can the student choose the plants that will be subjected to Environment B? **35 ways**

16. Holly wants to choose 5 different decorative tiles out of 8. If she plans to place the 5 tiles in a row, end to end, in how many different ways can she arrange them, from left to right? **6720 ways**

LESSON 11-1

Reteach

Permutations and Combinations

A **permutation** is a selection of items from a group in which the order is important. In a permutation, *AB* is NOT the same as *BA*.

The number of permutations of *n* items taken *r* at a time is shown by the following formula.

$${}_nP_r = \frac{n!}{(n-r)!}$$

The value of *r* must be less than or equal to the value of *n*.

How many ways can club members select a president, a vice president, a secretary, and a treasurer from a group of 10 members? Order matters since each office is different.

To find the number of permutations of 10 items taken 4 at a time, use $n = 10$ and $r = 4$ in the permutation rule. Then evaluate.

$${}_{10}P_4 = \frac{10!}{(10-4)!} = \frac{10!}{6!} = \frac{10\cdot9\cdot8\cdot7\cdot\cancel{6\cdot5\cdot4\cdot3\cdot2\cdot1}}{\cancel{6\cdot5\cdot4\cdot3\cdot2\cdot1}} = 10\cdot9\cdot8\cdot7 = 5040$$

Remember that *n*! or "*n* factorial" means to find the product of the whole numbers from 1 to *n*.

There are 5040 ways to select the officers.

Evaluate.

1. $8!$ **40,320**

2. $5!$ **120**

3. $10!$ **3,628,800**

4. $\frac{6!}{3!}$ **120**

5. $\frac{9!}{4!}$ **15,120**

6. $\frac{15!}{14!}$ **15**

Solve.

7. How many ways can the letters from *A* through *H* be used to create 5-letter passwords is there are no repeated letters in a password?

a. Does the order of the letters matter in the password? **Yes**

b. How many letters are there from *A* through *H*? **8**

c. Find the number of permutations of 8 letters taken 5 at a time.

$${}_8P_5 = \frac{8!}{(8-5)!} = \mathbf{6720}$$

8. An editor has 4 different spaces to arrange articles in a magazine. He must choose from 6 articles. How many different arrangements are possible?

Write and evaluate the permutation rule to solve.

$${}_6P_4 = \frac{6!}{(6-4)!} = 360$$

LESSON 11-1

Reteach

Permutations and Combinations (continued)

A **combination** is a selection of items from a group in which the order is NOT important. In a combination, *AB* is the same as *BA*.

The number of combinations of *n* items taken *r* at a time is shown by the following formula.

$${}_nC_r = \frac{n!}{(n-r)!r!}$$

The combination rule is a modification of the permutation rule.

How many ways can club members select a committee of 4 people from a group of 10 members?
The order of the committee members does not matter. Selecting Tom and Pat is the same as selecting Pat and Tom.

To find the number of combinations of 10 items taken 4 at a time, use $n = 10$ and $r = 4$ in the combination rule. Then evaluate.

$${}_{10}C_4 = \frac{10!}{(10-4)!4!} = \frac{10!}{6!4!} = \frac{10\cdot9\cdot8\cdot7\cdot\cancel{6\cdot5\cdot4\cdot3\cdot2\cdot1}}{(\cancel{6\cdot5\cdot4\cdot3\cdot2\cdot1})(4\cdot3\cdot2\cdot1)}$$

$$= \frac{10\cdot{}^3\cancel{9}\cdot\cancel{8}\cdot7}{\cancel{4\cdot3\cdot2}\cdot1} = 10\cdot3\cdot7 = 210$$

Divide out common factors to simplify.

There are 210 ways to select the committee members.

Evaluate.

9. ${}_5C_2$ 10

10. ${}_7C_6$ 7

11. ${}_9C_4$ 126

Solve.

12. A pollster has the names of 8 people available to answer her questions. She must select 3 of the people to interview. How many ways can she select the respondents?

a. Does the order of the respondents matter? No

b. Find the number of combinations of 8 people taken 3 at a time.

$${}_8C_3 = \frac{8!}{(8-3)!3!} = 56$$

13. The track team has 6 runners in a race. The coach selects 2 runners to run in the first heat. In how many ways can the runners be selected? Write and evaluate the combination rule to solve.

$${}_6C_2 = \frac{6!}{(6-2)!2!} = 15$$

LESSON 11-1

Challenge

Permutations with Special Conditions

A *permutation* is an arrangement of objects in a specific order. Sometimes there are also other conditions that must be satisfied. In such cases, you should deal with the special conditions first.

Using the letters in the word *square*, how many 6-letter arrangements with no repetitions are possible if vowels and consonants alternate, beginning with a vowel?

Of the 6 letters in the word, 3 are vowels (*u*, *a*, *e*) and 3 are consonants (*s*, *q*, *r*).

Beginning with a vowel, every other slot is to be filled by a vowel. There are 3 such slots and 3 vowels to be arranged in them.

3 × ___ × 2 × ___ × 1 × ___

The remaining 3 slots have 3 consonants to be arranged in them.

3 × 3 × 2 × 2 × 1 × 1

Multiply to determine the total number of arrangements.

There are 36 possible arrangements.

The girls Amy, Ann, and Doris and the boys Al, Arnon, Bob, and Roy are in a nursery group. Determine the number of ways the children can be arranged in a line with the following conditions.

1. A girl is always at the head of the line. 2160

2. Roy is always at the head of the line. 720

3. A child whose name begins with *A* is always at the head of the line. 2880

4. A child whose name begins with *A* is always at the head and the rear of the line. 1440

The diamond suit from a standard deck of 52 playing cards is removed from the deck, shuffled, and laid out in a row. Determine the number of possible arrangements.

5. The first card is the ace. 479,001,600

6. The first card is a face card. 1,437,004,800

Use the digits 0, 1, 2, 3, 4 without repetition. Determine the number of ways to form each arrangement.

7. 3-digit numerals whose values are at least 100 48

8. 4-digit numerals whose values are at least 1000 and less than 4000 72

LESSON 11-1

Problem Solving

Permutations and Combinations

Rosalie is looking at locks. The label *combination lock* confuses her. She wonders about the number of possible permutations or combinations a lock can have.

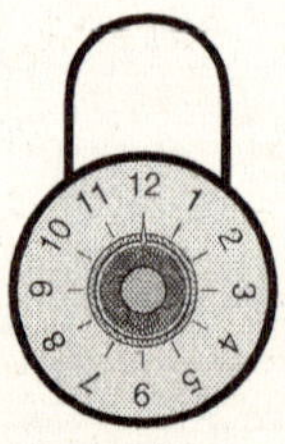

1. She looks at one circular lock with 12 positions. To open it she turns the dial clockwise to a first position, then counterclockwise to a second position, then clockwise to a third position

a. Write an expression for the number of 3-position codes that are possible, if no position is repeated.
12 × 11 × 10 = 1320

b. Explain how this represents a combination or a permutation.
Permutation; possible answer: the order of the 3 numbers matters.

2. Rosalie looks at cable locks. Each position can be set from 0 to 9. How many different codes are possible for each lock if no digits are repeated in each code?

a. a 3-digit cable lock
720 codes

b. a 4-digit cable lock
5040 codes

c. a 6-digit cable lock
151,200 codes

3. Rosalie needs 2 cable locks, but there are 13 types of locks to choose from.

a. In how many ways can she choose 2 different locks? 78 ways

b. Explain how this represents a permutation or a combination.
Combination; possible answer: the order in which she chooses the locks does not matter.

4. Explain why you think Rosalie might be confused by the label *combination lock*.
Because order matters, *combination locks* represent permutations.

Rosalie wants to lock her bicycle near the library. There are 7 slots still open in the bike rack. Choose the letter for the best answer.

5. Rosalie arrives at the same time as 2 other cyclists. In how many ways can they arrange their bikes in the open slots?
A 7
B 35
(C) 210
D 343

6. Suppose Rosalie arrived just ahead of the 2 other cyclists and selected a slot. In how many ways can the others arrange their bikes in the open slots?
F 2
G 15
H 24
(J) 30

LESSON 11-1

Reading Strategy

Compare and Contrast

Permutation	Combination
• Describes a grouping of objects in which order IS important	• Describes a grouping of objects in which order IS NOT important
• If John, Romeo, and Shelly all won prizes at the tennis tournament, it matters who won first prize, who won second prize, and who won third prize.	• If John, Romeo, and Shelly are chosen to compete in the tennis tournament, it doesn't matter in what order they are selected.
• The number of permutations of *x* items, chosen from a total of *n* items $_nP_x = \frac{n!}{(n-x)!}$	• The number of combinations of *x* items, chosen from a total of *n* items $_nC_x = \frac{n!}{x!(n-x)!}$
• *Example:* The number of ways that 3 objects can be selected from a total of 5 objects in a certain order. $_5P_3 = \frac{5!}{(5-3)!} = \frac{5\cdot4\cdot3\cdot2\cdot1}{2\cdot1} = 60$	• *Example:* The number of ways that 3 objects can be selected from a total of 5 objects in no particular order. $_5C_3 = \frac{5!}{3!(5-3)!} = \frac{5\cdot4\cdot3\cdot2\cdot1}{3\cdot2\cdot1(2\cdot1)} = 10$

Evaluate the following permutations and combinations.

1. ${}_4C_2$ 6

2. ${}_4P_2$ 12

3. ${}_5C_1$ 5

4. ${}_5P_1$ 5

5. ${}_3P_3$ 6

6. ${}_7C_6$ 7

Would you use a permutation or a combination to determine each of the following?

7. Nathan has a dozen eggs. He wants to decorate 4 eggs for an art project. Combination

8. Mr. Gunther is lining up 12 kindergarten students for a performance. Permutation

9. Rachel has 10 valuable baseball cards. She wants to select 2 of them to sell online. Combination

10. Don has 5 soccer trophies to line up on the mantel of his fireplace. Permutation

11. Bree has to select 5 photos from a box containing 25 photos to use in the yearbook. Combination

LESSON **11-2**

Practice A

Theoretical and Experimental Probability

Answer each question.

1. How many possible outcomes are there from tossing two number cubes labeled 1–6? **36 outcomes**
2. Describe the sample space for a spinner with four equal sections of blue, red, green, and yellow. **The sample space is blue, red, green, yellow.**
3. How likely is it that an outcome with a probability of 1 will occur? **Certain**
4. How likely is it that an outcome with a probability of 0 will occur? **Impossible**

Solve.

5. A farmer has four sheepdogs and three beagles. If he randomly chooses a dog to accompany him on a walk, what is the probability of him taking a walk with a sheepdog? $\frac{4}{7}$
6. Gordon spins a spinner with equal-sized sections numbered 1–6. In one spin, what is the likelihood that the spinner will stop on a 1 or a 5? $\frac{1}{3}$
7. Oak trees shade 30% of the Fitzgeralds' backyard. What is the probability that someone standing at a random point in the backyard will NOT be in the shade? $\frac{7}{10}$
8. Find the probability that a point chosen at random inside the larger square shown here will also fall inside the smaller square. $\frac{1}{9}$

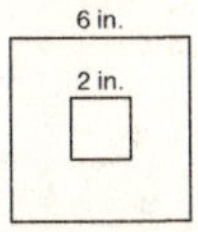

The table below shows the results of pulling one marble from a bag of marbles, recording its color, and replacing it in the bag.

Marble Color	Yellow	Red	Green
Times Pulled	53	17	30

Find the experimental probability of each event.

9. Choosing a yellow marble $\frac{53}{100}$
10. NOT choosing a red marble $\frac{83}{100}$
11. Choosing either a red or a green marble $\frac{47}{100}$
12. Which color marble is probably present in greatest number in the bag? **Yellow**

LESSON **11-2**

Practice B

Theoretical and Experimental Probability

Solve.

1. A fruit bowl contains 4 green apples and 7 red apples. What is the probability that a randomly selected apple will be green? $\frac{4}{11}$
2. When two number cubes labeled 1–6 are rolled, what is the probability that the result will be two 4's? $\frac{1}{36}$
3. Joanne is guessing which day in November is Bess's birthday. Joanne knows that Bess's birthday does not fall on an odd-numbered day. What is the probability that Joanne will guess the correct day on her first try? $\frac{1}{15}$
4. Tom has a dollar's worth of dimes and a dollar's worth of nickels in his pocket.
   a. What is the probability he will randomly select a nickel from his pocket? $\frac{2}{3}$
   b. What is the probability he will randomly select a dime from his pocket? $\frac{1}{3}$
5. Clarice has 7 new CDs; 3 are classical music and the rest are pop music. If she randomly grabs 3 CDs to listen to in the car on her way to school, what is the probability that she will select only classical music? $\frac{1}{35}$
6. Find the probability that a point chosen at random inside the larger circle shown here will also fall inside the smaller circle. $\frac{9}{16}$

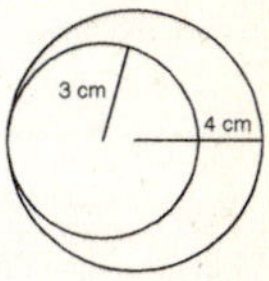

Frank is playing darts. The results of his throws are shown in the table below. Assume that his results continue to follow this trend.

Color Hit	Number of Throws
Blue	12
Red	5
White	2

Find the experimental probability of each event.

7. Frank's next throw will hit white. $\frac{2}{19}$
8. Frank's next throw will hit blue. $\frac{12}{19}$
9. Frank's next throw will hit either red or white. $\frac{7}{19}$
10. Frank's next throw will NOT hit red. $\frac{14}{19}$

LESSON **11-2**

Practice C

Theoretical and Experimental Probability

Solve.

1. A bowl contains 36 blue, 75 green, and 19 yellow jelly beans. What is the probability of randomly selecting a green jelly bean? $\frac{15}{26}$
2. Two spinners numbered 1–6 are spun. If all numbers are equally likely, what is the probability that the result will be two even numbers? $\frac{1}{4}$
3. Four quilters are preparing patches for a quilt. When finished, the quilt will contain 200 patches. The quilters' contributions thus far are in the table below.

Name	Number of Patches
Lia	65
Brian	17
Elle	88
Len	6

   a. What is the probability that a randomly chosen patch will have been sewn by Elle? $\frac{1}{2}$
   b. What is the probability that a randomly chosen patch will not have been sewn by Lia? $\frac{111}{176}$
   c. What is the probability that a randomly chosen patch will have been sewn by Brian or Len? $\frac{23}{176}$

A hacker is trying to break into his school's computer system to change his F's to A's. The computer system access password is 5 digits.

4. If digits in the password are allowed to repeat, what is the probability that the hacker will guess the password correctly on the first try? $\frac{1}{100,000}$
5. The hacker learns that the password does not contain any repeated digits. What is the new probability that he will randomly guess the password correctly? $\frac{1}{30,240}$
6. If the password contains no repeated digits, what is the probability that the digits in the school password have a sum less than 10? 0

Use the diagram to find each probability.

7. That a random point is within the circle in the triangle $\frac{\pi}{12}$
8. That a random point is NOT within the circle in the triangle $1 - \frac{\pi}{12}$

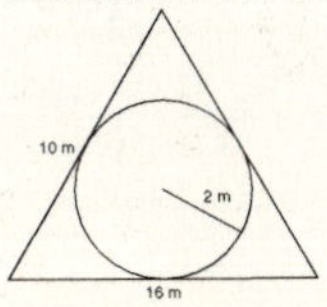

LESSON **11-2**

Reteach

Theoretical and Experimental Probability

Probability measures the likelihood of an event, or **outcome**. All of the possible outcomes make up the **sample space**.

For outcomes that have the same likelihood of occurring, or equally likely outcomes, the probability of an event, *E*, is defined as follows.

$$P(E) = \frac{\text{number of favorable outcomes}}{\text{number of outcomes in the sample space}}$$

$0 \le P(E) \le 1$

Two 1–6 number cubes are rolled. If the numbers on each cube are equally likely, what is the probability that the sum is 8?

The greatest possible sum is 12.

Step 1 Find the sample space.
Use ordered pairs to list all the possible outcomes.
There are 36 outcomes in the sample space.

6	(6, 1)	**(6, 2)**	(6, 3)	(6, 4)	(6, 5)	(6, 6)
5	(5, 1)	(5, 2)	**(5, 3)**	(5, 4)	(5, 5)	(5, 6)
4	(4, 1)	(4, 2)	(4, 3)	**(4, 4)**	(4, 5)	(4, 6)
3	(3, 1)	(3, 2)	(3, 3)	(3, 4)	**(3, 5)**	(3, 6)
2	(2, 1)	(2, 2)	(2, 3)	(2, 4)	(2, 5)	**(2, 6)**
1	(1, 1)	(1, 2)	(1, 3)	(1, 4)	(1, 5)	(1, 6)
	1	2	3	4	5	6

Rows show outcomes of the second number cube.

Columns show outcomes of the first number cube.

Step 2 Find the outcomes with a sum of 8:
(6, 2), (5, 3), (4, 4), (3, 5), and (2, 6).
There are 5 outcomes with a sum of 8.

These are the favorable outcomes.

Step 3 Use the definition of theoretical probability.

$$P(\text{sum is } 8) = \frac{\text{number of outcomes with sum of 8}}{\text{total possible outcomes}} = \frac{5}{36}$$

Two number cubes are rolled. If all numbers are equally likely, what is the probability of each event?

1. The sum is 3.
   a. List the favorable outcomes. **(2, 1), (1, 2)**
   b. Find the probability. $\frac{1}{18}$
2. The sum is 4.
   a. List the favorable outcomes. **(3, 1), (2, 2), (1, 3)**
   b. Find the probability. $\frac{1}{12}$
3. The sum is 9.
   a. List the favorable outcomes. **(6, 3), (5, 4), (4, 5), (3, 6)**
   b. Find the probability. $\frac{1}{9}$

LESSON 11-2 **Reteach**

Theoretical and Experimental Probability (continued)

To find the experimental probability of an event, conduct or observe an experiment for a large number of trials. Then count the number of times the event, A, occurs.

The **experimental probability** of the event is found by using the following formula.

$$P(A) = \frac{\text{number of times the event occurs}}{\text{number of trials}}$$

The table shows the results of an experiment with a spinner.

Spinner Experiment				
Color	Red	Yellow	Green	Blue
Spins	30	25	12	33

Outcomes

Trials

Find the experimental probability of spinning red.

Step 1 First find the total number of trials.
$30 + 25 + 12 + 33 = 100$

Step 2 Use the definition of experimental probability.

$$P(\text{red}) = \frac{\text{number of times red occurred}}{\text{number of trials}} = \frac{30}{100} = \frac{3}{10} = 0.3$$

You can use a fraction or a decimal.

To find the probability of spinning green or blue, find the total number of times green or blue occurred.

$$P(\text{green or blue}) = \frac{\text{number of times green or blue occurred}}{\text{number of trials}}$$
$$= \frac{12 + 33}{100} = \frac{45}{100} = \frac{9}{20} = 0.45$$

The table shows the results of a spinner experiment. Use the table to find each experimental probability.

Spinner Experiment				
Number	1	2	3	4
Spins	15	10	8	22

4. Spinning a 3
   a. What is the total number of trials? 55
   b. Find the probability. $\frac{8}{55}$
5. Spinning an even number
   a. What is the number of even spins? 32
   b. Find the probability. $\frac{32}{55}$
6. Spinning 1 or 2
   a. How many trials resulted in a 1 or a 2? 25
   b. Find the probability. $\frac{5}{11}$

LESSON 11-2 **Challenge**

Using Monte Carlo Methods to Determine Area

The first means of generating random numbers used equipment similar to that in gambling casinos, so methods of using random numbers to simulate events are called *Monte Carlo methods*, named after the casino in the principality of Monaco. These methods can be used to approximate area under a curve.

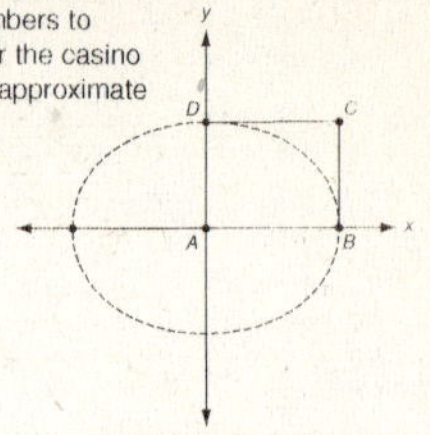

To approximate the area of the ellipse shown in the diagram, choose a point at random in rectangle $ABCD$. The probability that this point also lies in the first-quadrant region of the ellipse is the ratio of the area of the quarter of the ellipse to the area of rectangle $ABCD$. You can find this probability experimentally by generating R random points in rectangle $ABCD$ and counting the number of points, N, that also lie in the ellipse.

As shown at right, set the area ratio approximately equal to $\frac{N}{R}$, the probability that a point lies inside the ellipse.
$$\frac{\text{area of quarter of ellipse}}{\text{area of rectangle } ABCD} \approx \frac{N}{R}$$

Use the equation $\frac{x^2}{9} + \frac{y^2}{4} = 1$ for the ellipse.
$$\text{area of quarter of ellipse} \approx 6\left(\frac{N}{R}\right)$$

Substitute the area, 6, for rectangle $ABCD$ in the area ratio. Then multiply by 4 to determine the area of the entire ellipse.
$$\text{area of ellipse} \approx 24\left(\frac{N}{R}\right)$$

1. Use a graphing calculator to generate 25 points that lie in rectangle $ABCD$. Test whether these points lie inside the ellipse. Count the number of points, N, that are inside the ellipse.
   a. Evaluate your ratio.
   $\frac{N}{25}$ Possible answer: $\frac{17}{25} = 0.68$
   b. Approximate the area of the ellipse.
   Possible answer: 16.32 square units
   c. Use the formula πab to calculate the area of the ellipse.
   18.85 square units
   d. How can you refine your approximation?
   Increase the number of random points in a simulation. Repeat the simulations a number of times and take the average of the results.

Carry out a simulation to calculate the area. Repeat your simulation 2 times and take the average of the 3 results. Determine a theoretical value by using an area formula. Compare.

2. the area enclosed by the circle $x^2 + y^2 = 16$
   The area derived from the simulation will vary but should be close to 50.24 square units.
3. the area enclosed by the ellipse $\frac{y^2}{64} + \frac{x^2}{25} = 1$ above the x-axis
   The area derived from the simulation will vary but should be close to 62.8 square units

LESSON 11-2 **Problem Solving**

Theoretical and Experimental Probability

As part of a grant to improve bus routes to and from school, Hogan and Jane gather traffic flow statistics for one intersection. They make a table to show their findings for between 7:45 A.M. and 8:00 A.M. on a Monday morning.

Traffic Direction	Number of Vehicles
Straight through	282
Turn north	94
Turn south	188

1. Analyze the statistics.
   a. Write and evaluate an expression for $P(N)$, the probability that a vehicle will turn north.
   $P(N) = \frac{94}{564}$ 0.17
   b. Write and evaluate an expression for the probability that a vehicle will turn north or go straight through the intersection.
   $P(N \text{ or through}) = \frac{94 + 282}{564}$ 0.67
   c. Write and evaluate an expression for the probability that a vehicle will not turn north.
   $1 - P(N) = 1 - 0.17$ 0.83
2. The police department gathers statistics on Tuesday. Officers count a total of 608 vehicles, of which 380 go straight through the intersection, 76 turn north, and the rest turn south.
   a. What is the probability that a vehicle will turn north? $\frac{76}{608} = 0.125$
   b. What is the probability that a vehicle will turn north or go straight through the intersection? $\frac{76 + 380}{608} = 0.75$
   c. What is the probability that a vehicle will not turn north? $1 - 0.125 = 0.875$
3. Does this represent theoretical or experimental probability? Explain.
   Experimental; possible answer: the probabilities are based on actual data.

Math Assessment Survey				
Activity	Group projects	Keep a Journal	Multiple Choice	Word Problems
Student Response	57	18	35	10

A teacher surveys students on how they would prefer to have work assessed in math class. Choose the letter for the best answer.

4. What is the probability that a randomly chosen student prefers assessment through a group project?
   A $\frac{1}{12}$
   Ⓒ $\frac{19}{40}$
   B $\frac{53}{120}$
   D $\frac{21}{40}$
5. Which expression gives the probability that a randomly chosen student will not prefer multiple-choice questions?
   F $\frac{7}{24}$
   Ⓗ $1 - \frac{7}{24}$
   G $1 + \frac{7}{24}$
   J $35 - \frac{7}{24}$

LESSON 11-2 **Reading Strategy**

Understand Vocabulary

You can predict the likelihood that an event will take place by calculating the **probability** (P) of it occurring. In order to calculate the probability of an event occurring, you need to know all the possible **outcomes** of an event.

$$P = \frac{n}{t}$$

The number of favorable outcomes

The number of possible outcomes

If you toss a dime, there are 2 possible outcomes: heads and tails.

If you roll a number cube, there are 6 possible outcomes: 1, 2, 3, 4, 5, and 6.

If you draw a card from a full deck, there are 52 possible outcomes.

If you randomly choose one day of the week, there are 7 possible outcomes.

Identify n and t in each statement.

1. There are 3 black kittens in a litter of 6 kittens. $n = 3$, $t = 6$
2. In a box of 8 donuts, 2 are chocolate. $n = 2$, $t = 8$
3. Half the children in the 16-student class are girls. $n = 8$, $t = 16$
4. Harry put a silver dollar in 4 of the 12 grab bags. $n = 4$, $t = 12$

The **complement** of an event is the set of all unfavorable outcomes.

$$P(\text{not } E) = 1 - P(E)$$

$P(E) + P(\text{not } E) = 1$

Answer each question.

5. What is the complement to rolling a 3 on a number cube?
   Rolling a 1, 2, 4, 5, or 6
6. Fred has a bag containing 12 black marbles, 10 red marbles, and 15 white marbles.
   a. Fred wants to know the probability of randomly selecting a red marble from the bag. What are the unfavorable outcomes?
   Selecting a black marble or a white marble
   b. Write the formula to find the complement of choosing a red marble.
   $P(\text{not red}) = 1 - P(\text{red}) = 1 - \frac{10}{37} = \frac{27}{37}$

LESSON 11-3
Practice A
Independent and Dependent Events

Find each probability.

1. Hal is tossing a quarter.
   a. What is the probability he will toss heads? $\frac{1}{2}$
   b. What is the probability he will toss tails? $\frac{1}{2}$
   c. What is the probability he will toss heads and then tails? $\frac{1}{4}$
2. Hal tosses a quarter three times. What is the probability the result will be tails each time? $\frac{1}{8}$
3. Katie rolls a 1–6 number cube twice. What is the probability she will roll an odd number and then an even number? $\frac{1}{4}$
4. Katie rolls the 1–6 number cube three times. What is the probability that the result will be a 3 each time? $\frac{1}{216}$

There are 3 apples and 5 oranges in a bag. Determine each probability.

5. Selecting 2 apples when they are chosen at random without replacement $\frac{3}{28}$
6. Selecting an orange, then an apple when they are chosen at random without replacement $\frac{15}{56}$

A student must have a B average or better for all courses to qualify for any athletic team at Jefferson High School. The table below shows the distribution of students' grades in three sports at the school.

Sport	Students with an A Average	Students with a B Average
Field hockey	15	4
Basketball	7	13
Football	2	22

An athlete is randomly selected. Find each probability in decimal form.

7. The student is a field hockey player with a B average. 0.063
8. The student has an A average and plays football. 0.032
9. The student has a B average and does NOT play football. 0.27

There are 4 green marbles and 3 white marbles in a bag. A white marble is randomly selected and not replaced. Then a green marble is randomly selected.

10. Are these events dependent or independent? Dependent
11. What is the probability of this event occurring? $\frac{2}{7}$

LESSON 11-3
Practice B
Independent and Dependent Events

Find each probability.

1. A bag contains 5 red, 3 green, 4 blue, and 8 yellow marbles. Find the probability of randomly selecting a green marble, and then a yellow marble if the first marble is replaced. $\frac{3}{50}$
2. A sock drawer contains 5 rolled-up pairs of each color of socks, white, green, and blue. What is the probability of randomly selecting a pair of blue socks, replacing it, and then randomly selecting a pair of white socks? $\frac{1}{9}$

Two 1–6 number cubes are rolled—one is black and one is white.

3. The sum of the rolls is greater than or equal to 6 and the black cube shows a 3.
   a. Explain why the events are dependent. The events are dependent because $P(\text{sum} \geq 6)$ is different when it is known that a black 3 occurred.
   b. Find the probability. $\frac{1}{9}$
4. The white cube shows an even number, and the sum is 8.
   a. Explain why the events are dependent. The events are dependent because $P(\text{sum} = 8)$ is different when it is known that the white cube shows an even number.
   b. Find the probability. $\frac{1}{12}$

The table below shows numbers of registered voters by age in the United States in 2004 based on the census. Find each probability in decimal form.

Age	Registered Voters (in thousands)	Not Registered to Vote (in thousands)
18–24	14,334	13,474
25–44	49,371	32,763
45–64	51,659	19,355
65 and over	26,706	8,033

5. A randomly selected person is registered to vote, given that the person is between the ages of 18 and 24. 0.52
6. A randomly selected person is between the ages of 45 and 64 and is not registered to vote. 0.09
7. A randomly selected person is registered to vote and is at least 65 years old. 0.12

A bag contains 12 blue cubes, 12 red cubes, and 20 green cubes. Determine whether the events are independent or dependent, and find each probability.

8. A green cube and then a blue cube are chosen at random with replacement. Independent; $\frac{15}{121}$
9. Two blue cubes are chosen at random without replacement. Dependent; $\frac{3}{43}$

LESSON 11-3
Practice C
Independent and Dependent Events

Find each probability.

1. In cooking class, students are randomly choosing 1 of 3 different recipes. Two students choose the same recipe. $\frac{1}{9}$
2. Steven rolled a 1–6 number cube four times. The result was 4 odd numbers. $\frac{1}{16}$

The spinner shown here is spun twice.

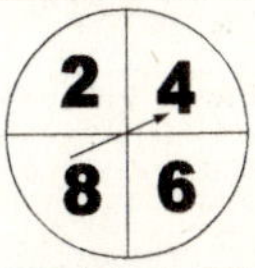

3. The sum of the results is equal to or greater than 10 and the first spin lands on 4.
   a. Explain why the events are dependent. Because $P(\text{sum} \geq 10)$ is different when it is known that the first spin lands on 4.
   b. Find the probability. $\frac{1}{8}$
4. The first spin lands on 6 and the sum of the results is less than or equal to 10.
   a. Explain why the events are dependent. Because $P(6)$ is different when it is known that the sum of both spins is less than or equal to 10.
   b. Find the probability. $\frac{1}{8}$

The table shows the population distribution in Ireland in 1996 by age and gender.

Ireland's Population in 1996

Age Group	0–20	21–40	41–60	61–80	Over 80
Males (in thousands)	620.4	526.8	405.3	212.0	33.0
Females (in thousands)	588.3	527.6	400.8	246.3	60.3

Find each probability.

5. A randomly selected person is no more than 20 years old, given that the person is male. 0.35
6. A randomly selected person is female, given that the person is over 80 years old. 0.65

A bag contains 3 red marbles, 7 yellow marbles, 5 green marbles, and 2 blue marbles. Determine whether the events are independent or dependent, and find each probability.

7. A red marble is chosen at random and replaced, and then the SAME red marble is chosen at random again. Independent; $\frac{3}{289}$
8. A yellow marble and then a blue marble are chosen at random without replacement. Dependent; $\frac{7}{136}$

LESSON 11-3
Reteach
Independent and Dependent Events

Two events, *A* and *B*, are **independent** if the occurrence of one does not affect the probability of the occurrence of the other.

Case 1: A card is drawn from a deck and then placed back in the deck. (Event A)
A second card is then drawn. (Event B)
Events *A* and *B* are independent.

Case 2: A card is drawn from a deck. It is not replaced. (Event C)
A second card is then drawn. (Event D)
Events *C* and *D* are NOT independent.

Multiplication Rule for the Probability of Independent Events

A and *B* are independent events.

$P(A \text{ and } B) = P(A) \cdot P(B)$

A deck of cards has 12 face cards and 40 number cards. A card is drawn from a deck and then placed back in the deck. A second card is then drawn. (The events are independent.)

What is the probability of drawing two face cards from the deck?

Step 1 Find the total number of cards. 12 + 40 = 52

Step 2 Find the probability of drawing a face card.

$P(\text{face card}) = \frac{12}{52} = \frac{3}{13}$

Step 3 Use the rule for the probability of independent events.

$P(2 \text{ face cards}) = P(\text{face card}) \cdot P(\text{face card})$

$= \frac{3}{13} \cdot \frac{3}{13} = \frac{9}{169}$

What is the probability of drawing a face card and then a number card from the deck?

$P(\text{number card}) = \frac{40}{52} = \frac{10}{13}$

$P(\text{face card, then number card}) = P(\text{face card}) \cdot P(\text{number card})$

$= \frac{3}{13} \cdot \frac{10}{13} = \frac{30}{169}$

Find each probability.

1. Ben rolls a 4 and then a 5 on a 1–6 number cube.
   a. $P(4)$ $\frac{1}{6}$
   b. $P(5)$ $\frac{1}{6}$
   c. $P(4, \text{then } 5) = P(4) \cdot P(5)$ $\frac{1}{36}$
2. Ben rolls a 3 and then an even number on a 1–6 number cube.
   a. $P(3)$ $\frac{1}{6}$
   b. $P(\text{even number})$ $\frac{1}{2}$
   c. $P(3) \cdot P(\text{even number})$ $\frac{1}{12}$

LESSON 11-3
Reteach
Independent and Dependent Events (continued)

Two events, *A* and *B*, are **dependent** if the occurrence of one affects the probability of the occurrence of the other.

A card is drawn from a deck. It is not replaced. — Event *A*
A second card is then drawn. — Event *B*
Events *A* and *B* are **dependent.**

The probability of *B* given that *A* has occurred is called the **conditional probability** of *B*, given *A*. It is denoted $P(B \mid A)$.

Multiplication Rule for the Probability of Dependent Events
A and *B* are dependent events.

$P(A \text{ and } B) = P(A) \cdot P(B \mid A)$ — The events are dependent.

A deck of cards has 12 face cards and 40 number cards. A card is drawn from a deck and NOT replaced. A second card is then drawn. What is the probability of drawing two face cards from the deck?

Step 1 Find the probability of drawing a face card on the first draw. — Event A
$P(\text{face card}) = \frac{12}{52} = \frac{3}{13}$

Step 2 Find the probability of drawing a face card on the second draw. — Event B
$P(\text{face card} \mid \text{face card}) = \frac{11}{51}$ — If 1 face card was drawn, 11 face cards remain and 51 total cards remain.

Step 3 Use the rule for the probability of dependent events.
$P(2 \text{ face cards}) = P(\text{face card}) \cdot P(\text{face card} \mid \text{face card})$
$= \frac{3}{13} \cdot \frac{11}{51} = \frac{33}{663} = \frac{11}{221}$

What is the probability of drawing a face card and then a number card from the deck?
$P(\text{number card} \mid \text{face card}) = \frac{40}{51}$ — There are still 40 number cards in the deck, but only 51 total cards remain.

$P(\text{face card, then number card}) = P(\text{face card}) \cdot P(\text{number card} \mid \text{face card})$
$= \frac{3}{13} \cdot \frac{40}{51} = \frac{40}{221}$

Find each probability.

3. Draw two number cards without replacing the first card.
   a. *P*(number card) $\frac{10}{13}$
   b. *P*(number card | number card) $\frac{13}{17}$
   c. *P*(number card) • *P*(number card | number card) $\frac{10}{17}$
4. Draw a number card and then a face card without replacing the first card.
   a. *P*(number card) $\frac{10}{13}$
   b. *P*(face card | number card) $\frac{4}{17}$
   c. *P*(number card) • *P*(face card | number card) $\frac{40}{221}$

LESSON 11-3
Challenge
Random Walks

The diagram at right shows a method of graphically recording the results of 6 coin tosses that occur in one-second intervals. The horizontal axis shows time, *t*, and the vertical axis shows position, *s*. Begin at the origin and move one unit up to record "heads" and one unit down to record "tails."

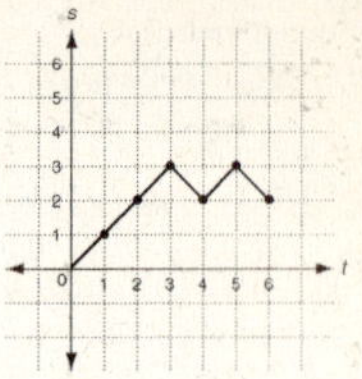

This type of graph is called a *random walk*.

1. Using *H* for "heads" and *T* for "tails," write the sequence for the random walk shown in the diagram at right.
   HHHTHT

Graph the random walk for the given coin sequences *A*, *B*, and *C*.

2. *A*: HHHHTTTHTHTTTTH
3. *B*: TTHTTHHTHTTHHHH
4. *C*: HHHHTTHTTHHHTTT

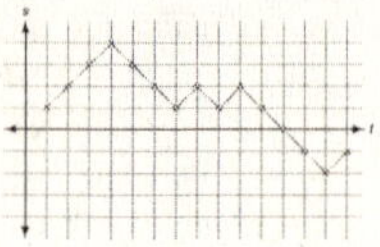
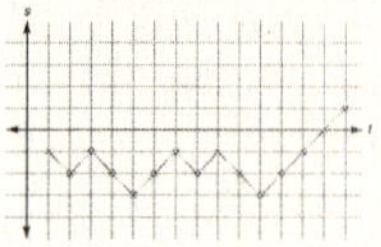
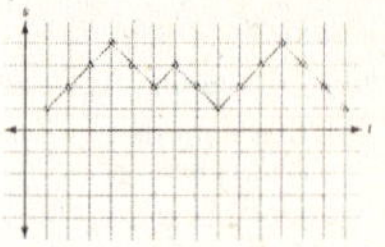

Refer to the graphs of sequences *A*, *B*, and *C*. For each sequence, give the time, if it exists, at which the random walk first returns to position $s = 0$. Then give the amount of time that the random walk spends in the first quadrant.

5. Sequence *A* — $t = 12$; 12 s
6. Sequence *B* — $t = 14$; 1 s
7. Sequence *C* — Does not exist; 15 s

Solve.

8. With a partner, toss a coin 20 times to generate a random walk. Generate 5 such walks.
   a. In what percent of your walks do you return to position $s = 0$ during the walk? Possible answer: 80%
   b. What is the average number of tosses to return to position $s = 0$? Possible answer: 6 tosses
9. Give an example of a real-life situation for which a random walk would be an appropriate model. The performance of a stock over time

LESSON 11-3
Problem Solving
Independent and Dependent Events

The table shows student participation in different sports at a high school. Suppose a student is selected at random.

Sports Participation by Grade

	Track	Volleyball	Basketball	Tennis	No Sport
Grade 9	12	18	15	9	66
Grade 10	6	20	12	2	95
Grade 11	15	11	8	5	61
Grade 12	7	6	10	12	50

1. What is the probability that a student is in grade 10 and runs track?
   a. Find the probability that a student is in grade 10, *P*(10). $P(10) = \frac{135}{440}$
   b. Find the probability that a student runs track, given that the student is in grade 10, $P(Tr \mid 10)$, $P(Tr \mid 10) = \frac{6}{135}$
   c. Find $P(10 \text{ and } Tr) = P(10) \cdot P(Tr \mid 10)$. $P(10 \text{ and } Tr) = \frac{6}{440} \approx 0.014$
2. What is the probability that a student is in grade 12 and runs track or plays tennis?
   a. Find the probability that a student is in grade 12, *P*(12). $P(12) = \frac{85}{440}$
   b. Find the probability that a student runs track or plays tennis, given that the student is in grade 12, $P(Tr \text{ or } Te \mid 12)$. $P(Tr \text{ or } Te \mid 12) = \frac{7+12}{85} = \frac{19}{85}$
   c. Find $P(12 \text{ or } (Tr \text{ or } Te))$. $P(12 \text{ and } (Tr \text{ or } Te)) = \frac{19}{85} \cdot \frac{85}{440} \approx 0.043$
3. During a fire drill, the students are waiting in the parking lot. What is the probability that one student is in grade 12 and runs track or plays tennis, and the student standing next to her is in grade 10 and runs track?
   a. Find the probability for the first student. $\frac{19}{440}$
   b. Find the probability for the second student. $\frac{6}{439}$
   c. Find the probability for the event occurring. $\frac{19}{440} \cdot \frac{6}{439} \approx 0.0006$
   d. Are these events independent or dependent? Explain.
   Dependent; possible answer: the second student is one of the remaining 439 students.

Samantha is 1 of 17 students in a class of 85 who have decided to pursue a business degree. Each week, a student in the class is randomly selected to tutor younger students. Choose the letter for the best answer.

4. What is the probability of drawing a business student one week, replacing the name, and drawing the same name the next week?
   A 3.4
   B 0.2
   C 0.04
   (D) 0.002
5. What is the probability of drawing Samantha's name one week, not replacing her name, and drawing the name of another business student the next week?
   (F) $\frac{1}{85} \cdot \frac{16}{84}$
   G $\frac{1}{85} \cdot \frac{17}{84}$
   H $\frac{17}{85} \cdot \frac{16}{84}$
   J $\frac{17}{85} \cdot \frac{17}{84}$

LESSON 11-3
Reading Strategy
Draw Conclusions

In determining probability, events can be independent or dependent.

Events are independent if the occurrence of one does not affect the probability of the other.	Events are dependent if the occurrence of one affects the probability of the other.
A bag contains 4 apples and 4 oranges. Choosing a piece of fruit, replacing it, and choosing again are **independent** events. In each case there is the same number of each type of fruit in the bag.	Choosing a piece of fruit and not replacing it changes the probability of what you will choose the second time. There are now only 7 pieces of fruit in the bag. These are **dependent** events.

In both cases, calculate the probability of two events by multiplying the probability of each event occurring. However, the calculation of the probability of the second event requires a conclusion as to whether the events are independent or dependent.

Determine if the events are dependent or independent.

1. Tossing heads 3 times in a row — Independent
2. Rolling a 5 on a number cube and then rolling another 5 — Independent
3. Choosing a red candy from a bag of multicolored candies, not replacing it, and then choosing a green candy from the bag — Dependent
4. Spinning a spinner and having it land on an odd number, and spinning it again and having it land on an even number — Independent
5. Drawing a red 8 from a deck of cards, replacing it, and then drawing a black queen — Independent
6. Drawing a face card from a deck of cards, not replacing it, and then drawing an ace — Dependent
7. Choosing a senior from the track team to run in the state relay race, and then choosing a junior from the track team to run in the local 10K race — Dependent

Which is the more likely event?

8. Event 1: Choosing a green marble from a bag containing 3 green marbles and 2 blue marbles
   Event 2: Rolling a 4 on a number cube
   Event 1
9. Event 1: Choosing the black crayon from a box of 20 crayons, replacing it, and choosing the white crayon
   Event 2: Choosing the black crayon from a box of 20 crayons, not replacing it, and choosing the white crayon
   Event 2

LESSON 11-4

Practice A
Compound Events

Determine which event(s) are mutually exclusive.

1. Students are forming 4 softball teams by each picking one of four different-color cards from a bag. Are the events "choosing a red card" and "choosing a blue card" mutually exclusive? Explain why or why not.

   These events are mutually exclusive since each student can choose only one card.

2. Are the events "choosing a black card" from a deck of playing cards and "choosing a 10" mutually exclusive? Explain why or why not.

   These events are not mutually exclusive since a card can be both black and a 10.

Solve.

3. Apples are in $\frac{1}{4}$ of all of the lunch bags that are distributed at a school picnic, and bananas are in $\frac{1}{3}$ of the bags. What is the probability of randomly choosing a lunch bag that contains either an apple or a banana? $\frac{7}{12}$

4. A group of senior citizens have won free vacation packages. The vacation to Bermuda is chosen by 25% of them, 60% choose Alaska, and 15% choose Costa Rica. What is the probability that one randomly selected senior citizen chooses to vacation in Costa Rica or Bermuda? $\frac{2}{5}$

Pam rolls a 1–6 number cube. Find each probability.

5. Pam rolls a 3 or a 6. $\frac{1}{3}$
6. Pam rolls an even number. $\frac{1}{2}$
7. Pam rolls an odd or even number. 1
8. Pam rolls an odd number or a 3. $\frac{1}{2}$
9. Pam rolls an odd number or a number greater than 4. $\frac{2}{3}$

Use the data to fill in the Venn diagram. Then solve.

10. Of the 65 students going on the soccer trip, 43 are players and 12 are left-handed. Only 5 of the left-handed students are soccer players. What is the probability that one of the students on the trip is a soccer player or is left-handed? $\frac{10}{13}$ or 0.77

LESSON 11-4

Practice B
Compound Events

A can of vegetables with no label has a $\frac{1}{8}$ chance of being green beans and a $\frac{1}{5}$ chance of being corn.

1. Explain why the events "green beans" or "corn" are mutually exclusive.

   These events are mutually exclusive because each can contains only one type of vegetable.

2. What is the probability that an unlabeled can of vegetables is either green beans or corn? $\frac{13}{40}$

Ben rolls a 1–6 number cube. Find each probability.

3. Ben rolls a 3 or a 4. $\frac{1}{3}$
4. Ben rolls a number greater than 2 or an even number. $\frac{5}{6}$
5. Ben rolls a prime number or an odd number. $\frac{2}{3}$

Of the 400 doctors who attended a conference, 240 practiced family medicine and 130 were from countries outside the United States. One-third of the family medicine practitioners were not from the United States.

6. What is the probability that a doctor practices family medicine or is from the United States? $\frac{7}{8}$
7. What is the probability that a doctor practices family medicine or is not from the United States? $\frac{29}{40}$
8. What is the probability that a doctor does not practice family medicine or is from the United States? $\frac{4}{5}$

Use the data to fill in the Venn diagram. Then solve.

9. Of the 220 people who came into the Italian deli on Friday, 104 bought pizza and 82 used a credit card. Half of the people who bought pizza used a credit card. What is the probability that a customer bought pizza or used a credit card? $\frac{67}{110}$ or 0.61

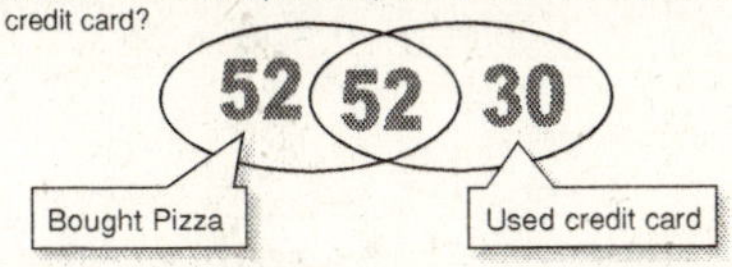

Solve.

10. There are 6 people in a gardening club. Each gardener orders seeds from a list of 11 different types of seeds available. What is the probability that 2 gardeners will order the same type of seeds? 0.81

LESSON 11-4

Practice C
Compound Events

Cards numbered 1–25 are placed in a bag and one is drawn at random. Find each probability.

1. Drawing an odd number or a multiple of 7. $\frac{14}{25}$
2. Drawing an even number or a perfect square. $\frac{3}{5}$

A drug company is testing the side effects of different doses of a new drug on three different groups of volunteers.

Group	Volunteers	Daily Amount (mg)
A	353	150
B	467	200
C	310	250

3. If a volunteer is chosen randomly, what is the probability that this person receives the highest amount per day? $\frac{31}{113}$ or 0.27
4. If a volunteer is chosen randomly, what is the probability that this person receives more than 150 mg per day? $\frac{777}{1130}$ or 0.69
5. If a volunteer is chosen randomly, what is the probability that this person does NOT receive 200 mg per day? $\frac{663}{1130}$ or 0.59

Mr. Rodney's English class is made up of 28 students. He has 6 ESL students, 10 remedial students, and 5 advanced learners. ESL students make up $\frac{1}{5}$ of the remedial students and $\frac{3}{5}$ of the advanced learners.

6. What is the probability that a student is ESL and remedial? 0.07
7. What is the probability that a student is ESL and an advanced learner? 0.11
8. What is the probability that a student is remedial and NOT ESL? 0.29

Solve.

9. A student is collecting a population of laboratory mice to be used in an experiment. He finds that of the 236 mice in the lab, 173 mice are female and 99 have pink eyes. Just 10 of the pink-eyed mice are male. What is the probability that a mouse is female or has pink eyes?

   $\frac{183}{236}$ or 0.78

10. A group of 4 friends buys a CD of computer screen savers. If there are a total of 12 screen savers on the CD, what is the probability that at least 2 of the friends will choose the same screen saver for their computer?

    0.43

LESSON 11-4

Reteach
Compound Events

A **simple event** is an event that cannot be broken down any further.
A **compound event** is an event made up of two or more simple events.

Rolling a 6 on a number cube is a simple event, while rolling an even number is a compound event.

Mutually exclusive events are events that cannot occur at the same time.

Tossing heads and tossing tails are mutually exclusive events. They cannot happen on the same toss of a coin.

Addition Rule for the Probability of Mutually Exclusive Events
A and *B* are mutually exclusive events.

$$P(A \text{ or } B) = P(A) + P(B)$$

A card is drawn from the deck. What is the probability of drawing a 5 or a face card from the deck?

Step 1 Decide whether the events are mutually exclusive.
A card cannot be both 5 and a face card.
The events are mutually exclusive.

Step 2 Find the probability of drawing a 5.
$P(5) = \frac{4}{52} = \frac{1}{13}$

There are four 5's in a deck of 52 cards.

Step 3 Find the probability of drawing a face card.
$P(\text{face card}) = \frac{12}{52} = \frac{3}{13}$

There are 12 face cards in a deck of 52 cards.

Step 4 Use the rule for the probability of mutually exclusive events.
$P(5 \text{ or face card}) = P(5) + P(\text{face card})$
$= \frac{1}{13} + \frac{3}{13} = \frac{4}{13}$

Toni draws a card from a deck. Find each probability.

1. She draws a number card or a king.
   a. *P*(number card) $\frac{10}{13}$
   b. *P*(king) $\frac{1}{13}$
   c. *P*(number card or king) = *P*(number card) + *P*(king) $\frac{11}{13}$
2. She draws an ace or a queen.
   a. *P*(ace) $\frac{1}{13}$
   b. *P*(queen) $\frac{1}{13}$
   c. *P*(ace or queen) = *P*(ace) + *P*(queen) $\frac{2}{13}$

LESSON 11-4

Reteach

Compound Events (continued)

Inclusive events are compound events that can occur at the same time.

Rolling a 6 and rolling an even number on a number cube are inclusive events because 6 is an even number.

Because inclusive events can happen simultaneously, you do not want to double count the probability when the events overlap.

Addition Rule for the Probability of Inclusive Events

A and *B* are inclusive events.

$P(A \text{ or } B) = P(A) + P(B) - P(A \text{ and } B)$

$P(A \text{ and } B)$ denotes the probability that both *A* and *B* occur at the same time.

A card is drawn from the deck. What is the probability of drawing a 3 or a heart from the deck?

Step 1 Decide whether the events are inclusive.
The events overlap: there is a card that is both a 3 and a heart.
The events are inclusive.

Step 2 Find the probability of drawing a 3. — There are four 3's in a deck of 52 cards.
$P(3) = \frac{4}{52}$

Step 3 Find the probability of drawing a heart. — There are 13 hearts in a deck of 52 cards.
$P(\text{heart}) = \frac{13}{52}$

Step 4 Find the probability of drawing a 3 and a heart. — There is 1 card that is both a 3 and a heart.
$P(3 \text{ and heart}) = \frac{1}{52}$

Step 5 Use the rule for the probability of inclusive events.
$P(3 \text{ or heart}) = P(3) + P(\text{heart}) - P(3 \text{ and heart})$
$= \frac{4}{52} + \frac{13}{52} - \frac{1}{52} = \frac{16}{52} = \frac{4}{13}$

Charles draws a card from a deck. Find each probability.

3. He draws a red card or a king.
   a. $P(\text{red})$ $\frac{26}{52}$
   b. $P(\text{king})$ $\frac{4}{52}$
   c. $P(\text{red and king})$ $\frac{2}{52}$
   d. $P(\text{red or king}) = P(\text{red}) + P(\text{king}) - P(\text{red and king})$ $\frac{7}{13}$
4. He draws a diamond or a face card.
   a. $P(\text{diamond})$ $\frac{13}{52}$
   b. $P(\text{face card})$ $\frac{12}{52}$
   c. $P(\text{diamond and face card})$ $\frac{3}{52}$
   d. $P(\text{diamond or face card}) = P(\text{diamond}) + P(\text{face card}) - P(\text{diamond and face card})$ $\frac{11}{26}$

LESSON 11-4

Challenge

Exploring P(A or B)

If *A* and *B* represent two events in the same sample space, then
$P(A \text{ or } B) = P(A) + P(B) - P(A \text{ and } B)$ when *A* and *B* are inclusive events.
When *A* and *B* are mutually exclusive events, this simplifies to $P(A \text{ or } B) = P(A) + P(B)$.
This formula can also be extended to apply to probabilities involving three events.

Consider the following situation.

A certain drug causes a skin rash or hair loss in 35% of patients. Twenty-five percent of patients experience only a skin rash, and 5% experience both a skin rash and hair loss. Find the probability that a patient will experience hair loss only.

1. Using *A* to represent "experiences a skin rash" and *B* to represent "experiences hair loss," write a symbolic representation of the problem.
   $P(B) = P(A \text{ or } B) - P(A) + P(A \text{ and } B)$
2. Using the symbolic representation you wrote, write the answer to the question. 15%

Consider the following situation.

Tawana and Sergio recorded a compact disc together. Each sang some solos, and the two sang some duets. Sergio recorded twice as many duets as solos, and Tawana recorded six more solos than duets. When a CD player selects one of these songs at random, the probability that it will select a duet is 25%.

3. Let *s* represent the number of solos that Sergio recorded.
   a. Write an equation to express the probability of randomly selecting a duet in terms of *s*. $\frac{1}{4} = \frac{2s}{5s + 6}$
   b. Solve the equation. Then determine the total number of songs recorded. 16 songs
   c. Find the probability of selecting a solo by Tawana or a duet when a CD player selects one song at random. $\frac{7}{8}$
4. Explain how the shaded area in the diagram shown below is used in the formula for $P(A \text{ or } B)$.

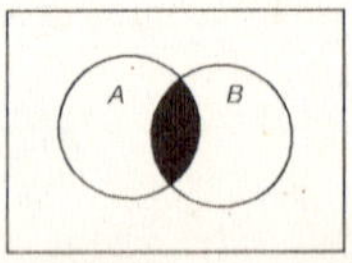

The probability of the shaded area is subtracted from the sum of the probabilities of the individual sets.

5. The diagram below represents three events in the same sample space. Use this diagram to write a formula for $P(A \text{ or } B \text{ or } C)$.

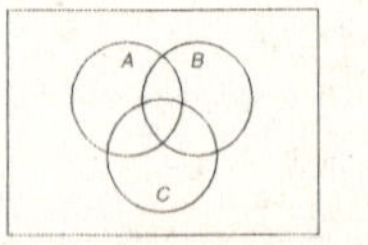

$P(A \text{ or } B \text{ or } C) = P(A) + P(B) + P(C) - P(A \text{ and } B) - P(A \text{ and } C) - P(B \text{ and } C) + P(A \text{ and } B \text{ and } C)$

LESSON 11-4

Problem Solving

Compound Events

Of 100 students surveyed, 44 are male and 54 are in favor of a change to a 9-period, 4-day school week. Of those in favor, 20 are female. One student is picked at random from those surveyed.

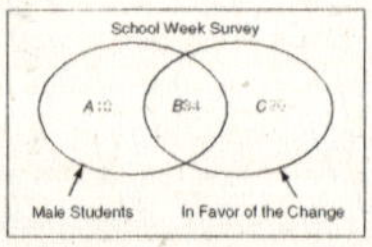

1. What is the probability that the student is male or favors the change? Use the Venn diagram.
   a. What is represented by the total of *A* + *B*?
      The total number of male students; 44
   b. What is represented by the total of *B* + *C*?
      The total number of students in favor of the change; 54
   c. How many of those in favor of the change are male? $54 - 20 = 34$
   d. Find the values for *A*, *B*, and *C* and label the diagram.
   e. Write and evaluate an expression for the probability that the student is male or favors the change. $\frac{44}{100} + \frac{54}{100} - \frac{34}{100} = \frac{64}{100} = 0.64$
2. What is the probability that the student is female or opposes the change?
   a. How many students are female? $100 - 44 = 56$
   b. How many students oppose the change? $100 - 54 = 46$
   c. If you draw a Venn diagram to show females and those opposed to the change, what is the meaning and value of the overlapping area?
      The number of females who are opposed to the change; 36
   d. Write and evaluate an expression for the probability that the student is female or opposes the change. $\frac{56}{100} + \frac{46}{100} - \frac{36}{100} = \frac{66}{100} = 0.66$
3. Of the students surveyed, 27 plan to start their own businesses. Of those, 18 are in favor of the change to the school week. Write and evaluate an expression for the probability that a student selected at random plans to start his or her own business or favors the change.
   $\frac{27}{100} + \frac{54}{100} - \frac{18}{100} = \frac{63}{100} = 0.63$

Sean asks each student to cast a vote for the type of class he or she would prefer. Of the students, 55% voted for online classes, 30% voted for projects, and 15% voted for following the textbook. Choose the letter for the best answer.

4. Which description best describes Sean's experiment?
   A Simple events
   B Compound events
   (C) Mutually exclusive events
   D Inclusive events
5. What is the probability that a randomly selected student voted for online classes or projects?
   F $\frac{33}{200}$
   G $\frac{7}{10}$
   H $\frac{1}{4}$
   (J) $\frac{17}{20}$

LESSON 11-4

Reading Strategy

Use a Venn Diagram

A Venn diagram can be used to organize what you know about mutually exclusive and inclusive events. The diagram on the left shows mutually exclusive events. There is no overlap so they can never occur at the same time. In the diagram on the right, the shaded overlapping region represents where both events A and B occur.

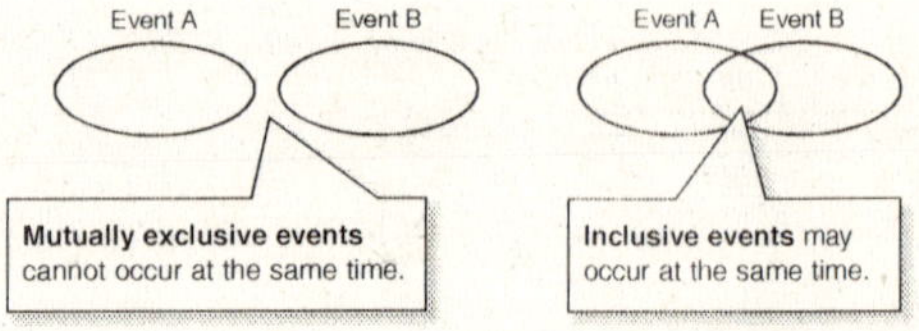

What is the probability of drawing a red card or a face card from a full deck?

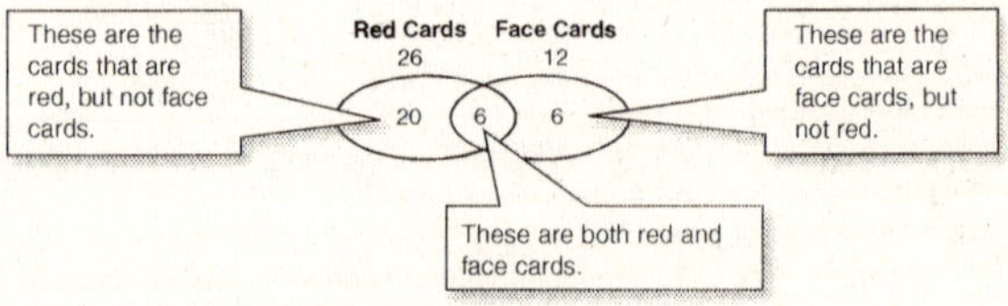

These are inclusive events because a card can be both red and a face card. Half of a deck of cards is red, so there are 26 red cards. There are 12 face cards in a deck. But some of the cards are both red and face cards. Of the 12 face cards, 6 of them are red. So, the overlapping region has a value of 6, since there are 6 cards that are favorable outcomes for both events A and B.

Draw and label a Venn diagram for each situation.

1. Of the 2250 voters surveyed, 1580 voted in the last election and 630 are planning to vote for Miller for mayor. Only 150 of those who voted in the last election are planning to vote for Miller.

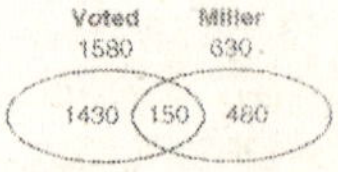

2. A restaurant served 253 customers last night. The staff reported that 189 of the customers had made a reservation and that 163 customers used credit cards. Only 47 of those who used credit cards did not have a reservation.

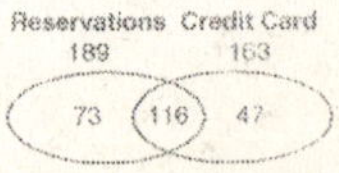

LESSON 11-5 Practice A
Measures of Central Tendency and Variation

Find the mean, median, and mode of each data set.

1. {3, 10, 2, 8, 7, 5, 2, 5}
   a. The mean is the sum of the values in a set divided by the number of values. Find the mean of the set. **5.25**
   b. The median is the middle value when the set is ordered numerically. Find the median of the set. **5**
   c. The mode is the value or values that occur most often. Find the mode of the set. **2, 5**
2. {11, 15, 4, 10, 7, 5, 11, 9}
   a. Mean **9** b. Median **9.5** c. Mode **11**

Find the expected values.

3. The probability distribution for the number of free throws that Larry makes in a game is given below. Find the expected number of free throws that Larry makes in a game. **6.07**

Number of Free Throws Made, n	5	6	7	8
Probability	0.21	0.58	0.14	0.07

4. The probability distribution for the number of pieces of junk mail May receives is given below. Find the expected number of junk mail letters May receives in a day. **2.05**

Number of Junk Mail Letters, n	1	2	3	4
Probability	0.15	0.70	0.10	0.05

Make a box-and-whisker plot of the data. Find the interquartile range.

5. {3, 7, 5, 3, 5, 9, 2, 7} **Interquartile range is 4.**

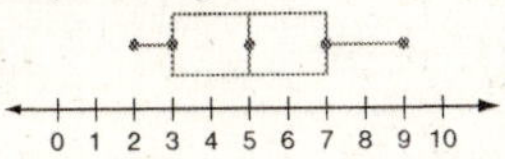

6. {1, 9, 9, 2, 2, 5, 5, 10} **Interquartile range is 7.**

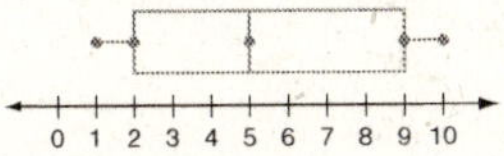

Find the variance and standard deviation.

7. {1, 2, 8, 11, 7, 10, 7, 2} **13; 3.6**
8. {10, 14, 8, 12, 9, 13} **4.7; 2.2**

LESSON 11-5 Practice B
Measures of Central Tendency and Variation

Find the mean, median, and mode of each data set.

1. {12, 11, 17, 3, 9, 14, 16, 2}
   a. Mean **10.5**
   b. Median **11.5**
   c. Mode **None**
2. {6, 9, 9, 20, 4, 5, 9, 13, 10, 1}
   a. Mean **8.6**
   b. Median **9**
   c. Mode **9**

Make a box-and-whisker plot of the data. Find the interquartile range.

3. {3, 7, 7, 3, 10, 1, 6, 6} **Interquartile range is 4.**

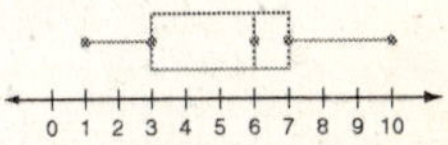

4. {1, 2, 3, 5, 3, 5, 8, 2} **Interquartile range is 3.**

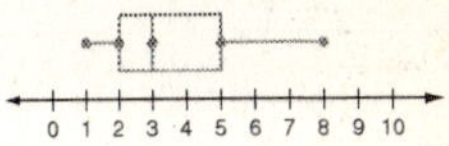

Find the variance and standard deviation.

5. {7, 4, 3, 9, 2} **6.8; 2.6**
6. {35, 67, 21, 16, 24, 51, 18, 32} **278; 16.7**
7. {19, 23, 17, 20, 25, 19, 15, 22} **9.3; 3.0**
8. {5, 12, 10, 13, 8, 11, 15, 12} **8.4; 2.9**

Solve.

9. The probability distribution for the amount of rain that falls on Boston in May each year is given below. Find the expected amount of rain for Boston in May. **7.01**

Inches of Rain, n	5	6	7	8
Probability	0.05	0.10	0.64	0.21

10. A biologist is growing bacteria in the lab. For a certain species of bacteria, she records these doubling times: 41 min, 45 min, 39 min, 42 min, 38 min, 88 min, 43 min, 40 min, 44 min, 39 min, 42 min, and 40 min.
    a. Find the mean of the data. **45.1**
    b. Find the standard deviation. **13.1**
    c. Identify any outliers. **88**
    d. Describe how any outlier affects the mean and the standard deviation. **The mean increases from ≈41.2 to ≈45.1, and the standard deviation increases from ≈2.1 to ≈13.1.**

LESSON 11-5 Practice C
Measures of Central Tendency and Variation

Write a data set to satisfy the given conditions.

1. Median = 8; mode = 4 **Possible answer: {4, 4, 8, 9, 10}**
2. Mean = 10; median = 12 **Possible answer: {3, 6, 12, 14, 15}**

Make a box-and-whisker plot of the data. Find the interquartile range.

3. 21, 20, 4, 5, 5, 20, 20, 13, 1, 1, 13, 7, 13, 17, 7, 17, 9, 9 **Interquartile range is 12.**

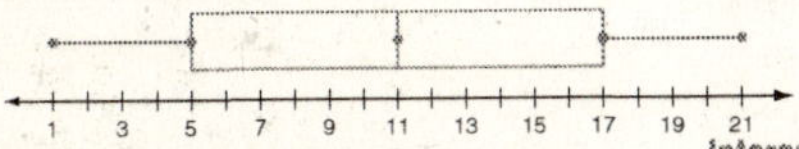

4. 10, 16, 3, 18, 18, 10, 8, 2, 9, 3, 7, 7, 8, 8, 18, 13, 7, 13 **Interquartile range is 6.**

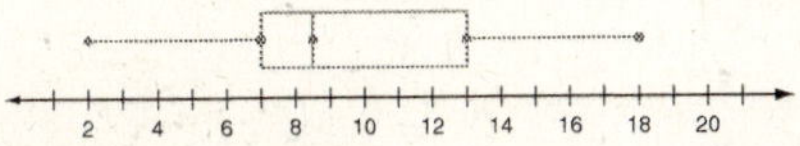

Find the variance and standard deviation.

5. {13, 7, 16, 22, 26, 11, 12, 19, 9} **35.1; 5.9**
6. {4, 7, 28, 6, 1, 1, 10, 15, 48, 3, 4, 5} **176.2; 13.3**
7. {15, 5, 12, 8, 19, 11, 7, 10, 9, 13, 17, 5} **18.6; 4.3**
8. {37, 29, 33, 30, 23, 28, 20, 35, 19, 21} **37.6; 6.1**

Solve.

9. The probability distribution for the number of children per family in a particular suburb of Chicago is shown below. Find the expected number of children per family in this region. **2.28**

Number of Children, n	1	2	3	4
Probability	0.27	0.28	0.35	0.10

10. A chemist weighs samples obtained from a production run. The weights of the samples are 13 g, 14 g, 65 g, 11 g, 15 g, 14 g, 14 g, 12 g, 13 g, 15 g, 14 g, and 12 g.
    a. Find the mean of the data. **17.7**
    b. Find the standard deviation. **14.3**
    c. Identify any outliers. **65**
    d. Describe how any outlier affects the mean and the standard deviation. **The mean increases from ≈13.4 to ≈17.7, and the standard deviation increases from ≈1.2 to ≈14.3.**

LESSON 11-5 Reteach
Measures of Central Tendency and Variation

Expected value can be thought of as the long-term average of an experiment after many trials.

The table shows a probability distribution for the variable x. A probability distribution table pairs each outcome with its probability.

Outcome	x_1	x_2	x_3	...	x_n
Probability	p_1	p_2	p_3	...	p_n

Expected value $= x_1p_1 + x_2p_2 + x_3p_3 + \dots + x_np_n$

> To find the expected value, find the sum of the products of each outcome and its associated probability.

This table shows the probability distribution for the number of red marbles selected when 3 marbles are randomly selected without replacement from a bag with 4 red and 6 blue marbles. Find the expected number of red marbles selected.

Outcome	0	1	2	3
Probability	0.17	0.5	0.3	0.03

In the formula for the expected value, $n = 4$.

Expected value $= x_1p_1 + x_2p_2 + x_3p_3 + x_4p_4$
$= 0(0.17) + 1(0.5) + 2(0.3) + 3(0.03)$
$= 0 + 0.5 + 0.6 + 0.09 = 1.19$

The expected number of red marbles is 1.19.

> Because the expected value is a type of average, it may not correspond exactly to any of the actual outcomes.

The probability distribution of temperatures, in degrees Celsius, from a chemistry experiment is shown below. Find the expected value of the temperature during the experiment.

Outcome	4	5	6	7	8
Probability	$\frac{2}{13}$	$\frac{4}{13}$	$\frac{3}{13}$	$\frac{4}{13}$	$\frac{2}{13}$

1. To use the formula for expected value, $n =$ **5**.
2. Write the expected value formula to use. **Expected value $= x_1p_1 + x_2p_2 + x_3p_3 + x_4p_4 + x_5p_5$**
3. Find the expected value of the temperature during the experiment. **≈6.9**

LESSON 11-5

Reteach

Measures of Central Tendency and Variation (continued)

The mean is a measure of central tendency. The variance and standard deviation are measures of dispersion or variation.
The notation often used for the mean is $\bar{x}$.
The variance is denoted σ^2 and the standard deviation is denoted σ.

This means they describe the spread of the data. The standard deviation measures spread from the mean.

Find the mean and standard deviation of the following **quiz scores**: 6, 8, 9, 8, 7, 5, 6, 7, 6, and 8.

Step 1 Find the mean.

$$\bar{x} = \frac{6+8+9+8+7+5+6+7+6+8}{10} = \frac{70}{10} = 7$$

Step 2 Find the difference between the mean and each data value.

Value of x	6	8	9	8	7	5	6	7	6	8
$x - \bar{x}$	−1	1	2	1	0	−2	−1	0	−1	1

Step 3 Square each value of $x - \bar{x}$.

$x - \bar{x}$	−1	1	2	1	0	−2	−1	0	−1	1
$(x - \bar{x})^2$	1	1	4	1	0	4	1	0	1	1

Step 4 Find the variance. Add all of the $(x - \bar{x})^2$ terms and divide by the number of data values.

$$\sigma^2 = \frac{1+1+4+1+0+4+1+0+1+1}{10} = \frac{14}{10} = 1.4$$

Step 5 Find the standard deviation.
The standard deviation is the square root of the variance.
$\sigma = \sqrt{1.4} \approx 1.18$

Dana's quiz scores are 2, 1, 5, 3, 1, 5, 3, and 4.

4. Find the mean. 3

5. Find the difference between the mean and each data value: $x - \bar{x}$.
−1, −2, 2, 0, −2, 2, 0, 1

6. Find $(x - \bar{x})^2$ for each data value.
1, 4, 4, 0, 4, 4, 0, 1

7. Find the variance.
2.25

8. Find the standard deviation.
1.5

LESSON 11-5

Challenge

Survey Data Analysis

Cassie's job is to analyze a set of 100 integral value scores that range from 0 to 10 from a recent survey. The set of values includes exactly 6 scores that are zero. She decides not to include the 6 zero scores in the analysis, based on the false assumption that zero is not really a number since it has no value.

1. For the set of 94 values, excluding the zeros, Cassie correctly computes the mean. How does the mean for the 94 nonzero values compare to the mean for the 100 values? Is it greater than, less than, equal to, or is it impossible to determine? How do you know?
Greater than; the sum is the same for both, and in the first case you divide by 94 rather than 100.

2. For the set of 94 values, excluding the zeros, Cassie correctly computes the median. How does the median for the 94 nonzero values compare to the median for the 100 values? Is it greater than, less than, equal to, or is it impossible to determine? How do you know?
Impossible to determine; it may be the same or it may be greater depending on the distribution of the data.

3. For the set of 94 values, excluding the zeros, Cassie correctly computes the mode. How does the mode for the 94 nonzero values compare to the mode for the 100 values? Is it greater than, less than, equal to, or is it impossible to determine? How do you know?
Equal to: There were 6 zero scores. At least one value has to occur more than 10 times since 94 divided by 9 is greater than 10.

4. For the set of 94 values, excluding the zeros, Cassie correctly computes the standard deviation. How does the standard deviation for the 94 nonzero values compare to the standard deviation for the 100 values? Is it greater than, less than, equal to, or impossible to determine? How do you know?
Less than; The set of 94 values ranges from 1 to 10 and so is not as spread out as the set of 100 values.

The frequency distribution for the 100 survey scores is shown below.

Survey Value	0	1	2	3	4	5	6	7	8	9	10
Frequency	6	8	9	11	13		12	10	8	6	5

5. How many people answered with the value of 5 in the survey? How do you know?
12; the sum of the set of frequencies must equal 100.

6. If you randomly chose 1 of the 100 people surveyed, what answer would you expect that person to give?
The expected value is 4.81, which gives an expected result of 5.

7. Find the first and third quartiles and the interquartile range for this set of values.
Median is 5, Q1 is 3, Q3 is 7, and interquartile range is 4.

LESSON 11-5

Problem Solving

Measures of Central Tendency and Variation

Each week, Damien records the miles per gallon for his car, to the nearest whole number. Over a period of 10 weeks, the data are 18, 17, 19, 18, 25, 29, 30, 26, 19. He wants to arrange and summarize his data so that he can analyze it.

1. Make a box-and-whisker plot of his data.
   a. Order the data from least to greatest. 17, 18, 18, 18, 19, 19, 25, 26, 29, 30
   b. Identify the minimum, maximum, median, first quartile, and third quartile.
   Minimum = 17; maximum = 30; median = 19; first quartile = 18; third quartile = 26
   c. Use the number line to make a box-and-whisker plot of the data. Find and label the interquartile range.

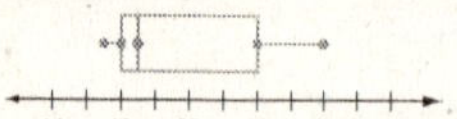

   d. Explain what the interquartile range represents in terms of the car's miles per gallon.
   Possible answer: Damien's car gets between 18 to 26 miles per gallon 50% of the time.

2. Find the standard deviation for the data.
   a. Write an equation and solve to find the mean. (18 + 17 + 19 + 18 + 18 + 25 + 29 + 30 + 26 + 19)/10 = 21.9
   b. Complete the table to show the difference between the mean and each data value, and the square of that difference.

Data Value, x	18	17	19	18	18	25	29	30	26	19
$x - \bar{x}$	−3.9	−4.9	−2.9	−3.9	−3.9	3.1	7.1	8.1	4.1	−2.9
$(x - \bar{x})^2$	15.21	24.01	8.41	15.21	15.21	9.61	50.41	65.61	16.81	8.41

   c. Explain how to use the data from the table to find the standard deviation.
   Find the square root of the mean of the $(x - \bar{x})^2$ terms.
   d. What is the standard deviation for the data? 4.78
   e. Explain what the standard deviation represents in terms of the car's miles per gallon.
   Possible answer: All but 3 data points are clustered within 1 standard deviation of the mean.

3. Damien thinks that the standard deviation is a more reliable measure of dispersion than the interquartile range. Is he correct? Explain.
He is correct; possible answer: the standard deviation depends on the entire set of data values; whereas the interquartile range depends on only 2 values, the first and third quartiles.

LESSON 11-5

Reading Strategy

Analyze Information

A box-and-whisker plot visually shows the distribution of data in a set. There are 9 members of the youth chess club. The following data set shows their ages.

{11, 17, 12, 16, 17, 14, 12, 9, 16}

The box-and-whisker plot of the data is shown below.

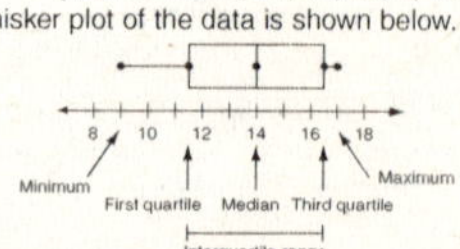

The median divides the data set into two parts, or quartiles. The median of the lower half of the data is the first quartile and the median of the upper half of the data is the third quartile. Where is the second quartile? It is the median of the full data set.

Answer each question.

1. The first quartile of some test scores is 55 and the interquartile range is 19.
   a. Explain how you can find the third quartile.
   Possible answer: Add the interquartile range to the first quartile. So the third quartile is 55 + 19 = 74.
   b. Is it possible for the median of the test scores to be 77? Explain why or why not.
   No; possible answer: the value of the median is between the first and third quartiles.
   c. The maximum score is 93 and the range of the test scores is 55. What is the minimum test score?
   93 − 55 = 38
   d. Which quantity describes the middle 50% of the test scores? What is the range of test scores that will place a student in the middle 50% of the class?
   The interquartile range; scores between 55 and 74

2. Which data set is represented by the box-and-whisker plot shown?
A = {7, 9, 6, 12, 9, 15, 8, 14, 12, 10}
B = {6, 9, 7, 11, 8, 15, 11, 14, 10, 13}
C = {7, 8, 8, 12, 9, 13, 11, 14, 14, 15}
B

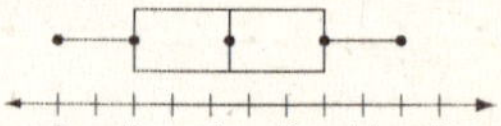

3. What do the "whiskers" in a box-and-whisker plot represent?
Possible answer: The whiskers are the lines drawn from the box to show the minimum and the maximum.

LESSON 11-6

Practice A
Binomial Distributions

Use the Binomial Theorem to expand each binomial.

1. $(x + y)^2$

   $x^2 + 2xy + y^2$

2. $(a + b)^3$

   $a^3 + 3a^2b + 3ab^2 + b^3$

3. $(2x + y)^2$

   $4x^2 + 4xy + y^2$

4. $(x + 3y)^3$

   $x^3 + 9x^2 + 27xy^2 + 27y^3$

Solve.

5. A coin lands heads half of the time it is tossed.

   a. What is the probability that the coin will land heads? 0.5

   b. What is the probability that the coin will NOT land heads? 0.5

   c. If the coin is tossed 5 times, what is the probability that it will land heads twice?

   0.31

   d. The probability that the coin will land heads either once or twice is given by $P(1 \text{ or } 2) = P(1) + P(2)$. Find the probability.

   0.47

6. A survey showed that 1 in 3 people entering the post office are wearing glasses. At noon, 4 people enter the post office.

   a. What is the probability that one of those people is wearing glasses?

   0.40

   b. What is the probability that one of those people is NOT wearing glasses?

   0.60

   c. What is the probability that at least 2 of the 4 people are wearing glasses?

   0.41

   d. The probability that at least 3 of the people are wearing glasses is given by $P(\geq 3) = P(3) + P(4)$. Find the probability.

   0.11

LESSON 11-6

Practice B
Binomial Distributions

Use the Binomial Theorem to expand each binomial.

1. $(x + y)^3$

   $x^3 + 3x^2y + 3xy^2 + y^3$

2. $(2x + y)^4$

   $16x^4 + 32x^3y + 24x^2y^2 + 8xy^3 + y^4$

3. $(m + 3n)^3$

   $m^3 + 9m^2n + 27mn^2 + 27n^3$

4. $(p + q)^5$

   $p^5 + 5p^4q + 10p^3q^2 + 10p^2q^3 + 5pq^4 + q^5$

Solve.

5. Of the new cars in a car dealer's lot, 1 in 6 are white. Today, 4 cars were sold.

   a. What is the probability that 3 of the cars sold were white?

   0.015

   b. What is the probability that at least 2 of the cars sold were white?

   0.13

6. At a small college, $\frac{1}{3}$ of all of the students are vegetarians. There are 5 students in line at the cafeteria.

   a. What is the probability that all 5 students are vegetarians?

   0.004

   b. What is the probability that just 1 of the students is a vegetarian?

   0.33

7. Ellen plays 8 hands of a card game with her friends. She has a 1 in 3 chance of winning each hand. What is the probability that she will win exactly half of the hands played?

   0.17

8. In a lottery, each ticket buyer has a 1 in 10 chance of winning a prize. If Chip buys 10 tickets, what is the probability that he will win at least 1 prize?

   0.65

LESSON 11-6

Practice C
Binomial Distributions

Use the Binomial Theorem to expand each binomial.

1. $(x + y)^5$

   $x^5 + 5x^4y + 10x^3y^2 + 10x^2y^3 + 5xy^4 + y^5$

2. $(4x + y)^4$

   $256x^4 + 256x^3y + 96x^2y^2 + 16xy^3 + y^4$

3. $(2x + y)^5$

   $32x^5 + 80x^4y + 80x^3y^2 + 40x^2y^3 + 10xy^4 + y^5$

4. $(n + 2m)^4$

   $16m^4 + 32m^3n + 24m^2n^2 + 8mn^3 + n^4$

Solve.

5. At Hopewell High School, 1 in 7 students is on a sports team. There are 4 student council representatives in the school.

   a. What is the probability that 2 of the student council representatives are also on a sports team?

   0.09

   b. What is the probability that at least 3 of the student council representatives are on a sports team?

   0.01

6. Two in every 5 donuts in a package of donuts have a jelly filling. There are 5 donuts left in the package.

   a. What is the probability that all 5 donuts have a jelly filling?

   0.01

   b. What is the probability that none of the donuts has a jelly filling?

   0.08

7. Andrew is choosing CDs from a bag of free CDs without looking. He has a 1 in 5 chance of choosing a CD that he likes. He chooses 8 CDs in all. What is the probability that he will get 3 CDs that he likes?

   0.15

8. In a game of Bingo, the contestants have a 1 in 12 chance of winning each round. If Shirley plays 6 rounds, what is the probability that she will win at least half of them?

   0.01

LESSON 11-6

Reteach
Binomial Distributions

The coefficients of the expansion of $(x + y)^n$ are the numbers in Pascal's triangle. The binomial coefficients are also combinations.

Pascal's Triangle	Combinations	Binomial Expansion
1	${}_0C_0$	$(x + y)^0 = 1$
1 1	${}_1C_0$ ${}_1C_1$	$(x + y)^1 = x + y$
1 2 1	${}_2C_0$ ${}_2C_1$ ${}_2C_2$	$(x + y)^2 = x^2 + 2xy + y^2$
1 3 3 1	${}_3C_0$ ${}_3C_1$ ${}_3C_2$ ${}_3C_3$	$(x + y)^3 = x^3 + 3x^2y + 3xy^2 + y^2$

The **Binomial Theorem** summarizes the relationship.

$$(x + y)^n = {}_nC_0x^ny^0 + {}_nC_1x^{n-1}y^1 + {}_nC_2x^{n-2}y^2 + \ldots + {}_nC_{n-1}x^1y^{n-1} + {}_nC_nx^0y^n$$

The sum of the exponents for each term is *n*.

You can use the Binomial Theorem to expand any binomial.

Expand $(x + y)^5$.

Step 1 Use the Binomial Theorem with n = 5.

$(x + y)^5 = {}_5C_0x^5y^0 + {}_5C_1x^4y^1 + {}_5C_2x^3y^2 + {}_5C_3x^2y^3 + {}_5C_4x^1y^4 + {}_5C_5x^0y^5$

Step 2 Use the fifth row of Pascal's triangle to find the coefficients. Continue the pattern and add pairs of terms from the fourth row to find the numbers in the fifth row.

```
      1    3    3    1          ← n = 3
   1    4    6    4    1        ← n = 4
1    5    10   10   5    1      ← n = 5
```

The sum of the exponents for each term is 5. The exponents of *x* decrease as the exponents of *y* increase.

Step 3 Write the coefficients.

$(x + y)^5 = 1x^5y^0 + 5x^4y^1 + 10x^3y^2 + 10x^2y^3 + 5x^1y^4 + 1x^0y^5$

Step 4 Simplify.

$(x + y)^5 = x^5 + 5x^4y + 10x^3y^2 + 10x^2y^3 + 5xy^4 + y^5$

Use the Binomial Theorem to expand each binomial.

1. $(2x + y)^3 = {}_3C_0(2x)^3y^0 + {}_3C_1(2x)^2y^1 + {}_3C_2(2x)^1y^2 + {}_3C_3(2x)^0y^3$

   $= {}_3C_0 \cdot 8x^3y^0 + {}_3C_1 \cdot 4x^2y^1 + {}_3C_2 \cdot 2xy^2 + {}_3C_3 \cdot y^3$

   $= 1 \cdot 8x^3 + 3 \cdot 4x^2y + 3 \cdot 2xy^2 + 1 \cdot y^3$

   $= 8x^3 + 12x^2y + 6xy^2 + y^3$

2. $(x + 3y)^4 = {}_4C_0x^4(3y)^0 + {}_4C_1x^3(3y)^1 + {}_4C_2x^2(3y)^2 + {}_4C_3x^1(3y)^3 + {}_4C_4x^0(3y)^4$

   $= 1 \cdot x^4 + 4 \cdot (3x^3y) + 6 \cdot (9x^2y^2) + 4 \cdot (27xy^3) + 1 \cdot (81y^4)$

   $= x^4 + 12x^3y + 54x^2y^2 + 108xy^3 + 81y^4$

LESSON 11-6

Reteach

Binomial Distributions (continued)

A binomial experiment must meet the following conditions.

- It has a fixed number of trials, n.
- The trials are independent.
- The outcomes of the trials must fall into two categories.
- The probabilities remain constant for each trial.

> The categories are classified as success and failure.

If a binomial experiment has n trials with probability of success p and probability of failure q for any trial, then the binomial probability of **exactly** r successes is given by the following formula.

$$P(r) = {}_nC_r p^r q^{n-r}$$

> Remember that $q = 1 - p$.

In a game with a number cube, a person wins if the number 1 or 2 is rolled. The person loses if the number 3, 4, 5, or 6 is rolled. In 4 rolls, find the probability of exactly 2 wins.

> The game satifies the conditions of a binomial experiment. There are 4 independent trials, each with fixed probability of success or failure.

Step 1 Find the probability of success (win) and failure (lose).

$p = \frac{2}{6} = \frac{1}{3}$

So $q = 1 - p = 1 - \frac{1}{3} = \frac{2}{3}$.

Step 2 Use the binomial probabilty formula with $r = 2$ successes, or wins, $n = 4$ rolls, or trials, $p = \frac{1}{3}$, and $q = \frac{2}{3}$.

$$P(2) = {}_4C_2\left(\frac{1}{3}\right)^2\left(\frac{2}{3}\right)^{4-2}$$
$$= {}_4C_2\left(\frac{1}{3}\right)^2\left(\frac{2}{3}\right)^2$$
$$= 6\left(\frac{1}{9}\right)\left(\frac{4}{9}\right)$$
$$= \frac{8}{27}$$

> Think: ${}_4C_2$ is the second term in the fourth row of Pascal's triangle.

So, the probability is $\frac{8}{27}$ that 2 out of 4 rolls of a number cube will result either a 1 or a 2.

Diane and Ed are playing a game with a number cube. Diane wins if the number 6 is rolled. She loses if any other number is rolled. Find the probability of Diane having exactly 2 wins in 3 rolls.

3. Find p and q. $p = \frac{1}{6}, q = \frac{5}{6}$
4. Find n and r. $n = 3, r = 2$
5. Write the appropriate binomial probability formula. $P(2) = {}_3C_2\left(\frac{1}{6}\right)^2\left(\frac{5}{6}\right)^{3-2}$
6. Find the probability. $\frac{5}{72}$

LESSON 11-6

Challenge

The Dice Game

Antoine Gombauld, the Chevalier de Méré, was a seventeenth century French dice player. He was a successful gambler in one game of dice rolling and unsuccessful in another. He would bet that he could roll at least one 6 in 4 rolls of a single die. With this bet he made money in the long run. But he would also bet that he could roll at least one pair of 6s in 24 rolls of a pair of dice. He eventually lost money with this bet.

The first experiment is rolling at least one 6 in 4 rolls of a single die.

1. Is each of the 4 rolls of the single die independent of one another? Why or why not?
   Yes; the probability of a 6 is always $\frac{1}{6}$.
2. What is the probability of rolling at least one 6 in 4 rolls? This means rolling either one 6, two 6s, three 6s, or four 6s.
   $P(\text{1 six}) \approx 0.39$; $P(\text{2 sixes}) \approx 0.12$; $P(\text{3 sixes}) \approx 0.015$; $P(\text{4 sixes}) \approx 0.000778$; $P(\text{at least 1 six}) \approx 0.53$
3. What is the probability of rolling no 6s in the 4 rolls? How does your answer compare to your previous answer?
   $P(\text{0 sixes}) \approx 0.487$; it is the complement so the sum of the probabilities is equal to 1.
4. Predict the outcome of Gombauld's bets with this game over time. How much might he win or lose for each \$1 bet?
   Possible answer: win over time; $(1)(0.53) + (-1)(0.47) = 0.06$, so he would win an average of 6 cents for a \$1 bet on every game.

Consider the game of rolling at least one pair of 6s in 24 rolls of a pair of dice.

5. Is each of the 24 rolls of the pair of dice independent? Why or why not?
   Yes; the probability of rolling a pair of 6s when rolling a pair of dice is always $\frac{1}{36}$.
6. What is the probability of rolling at least one pair of 6s in 24 rolls of a pair of dice? How can you find this probability?
   You can use the complement of rolling no pairs of 6s; $\approx 1 - 0.51 = 0.49$.
7. Predict the outcome of Gombauld's bets with this game over time. How much might he win or lose for each \$1 bet?
   Possible answer: lose over time; $(1)(0.49) + (-1)(0.51) = -0.02$, so he would lose an average of 2 cents for a \$1 bet on every game.
8. How might Antoine Gombauld have changed the rules so that he could have had a positive expected return for this game?
   Possible answer: He could have changed the rules so that he rolls at least one pair of 6s in 25 (or more) rolls of a pair of dice. This makes his expected value positive.

LESSON 11-6

Problem Solving

Binomial Distributions

Sales records for the snack machines show that 1 out of every 6 students buys a bag of trail mix. There are 5 students waiting to use the machines. Melanie uses the formula for binomial probability, $P(r) = {}_nC_r p^r q^{n-r}$, to determine the number of students expected to buy trail mix.

1. What is the probability of exactly 3 students buying a bag of trail mix?
   a. What is the probability of each student buying a bag of trail mix? $\frac{1}{6}$
   b. Define each variable used in the formula and give its value. $n = 5$; $r = 3$; $p = \frac{1}{6}$; $q = \frac{5}{6}$
   c. Write the binomial formula using these values. $P(3) = {}_5C_3\left(\frac{1}{6}\right)^3\left(\frac{5}{6}\right)^2$
   d. Solve the equation to give the probability of exactly 3 students buying a bag of trail mix. 0.032
2. What is the probability of at least 1 student buying a bag of trail mix?
   a. Describe a method to solve involving the sum of probabilities.
      Find the sum of the probabilities of 1, 2, 3, 4, and 5 students buying a bag of trail mix, $P(1) + P(2) + P(3) + P(4) + P(5)$.
   b. Describe a method to solve that uses the formula $P(E) + P(\text{not } E) = 1$.
      Find the probability that no student will buy a bag of trail mix and subtract that probability from 1.
   c. Use either method to determine the probability of at least 1 student buying a bag of trail mix. 0.6
3. After school, 4 students line up to buy snacks from the machine. What is the probability that they will all buy a bag of trail mix? 0.0008

Sports drinks are purchased by 3 out of 4 students using the snack machines. There are 3 students at the machines now. Choose the letter for the best answer.

4. Which expression gives the probability of exactly 2 students buying an energy drink?
   (A) $P(2) = {}_3C_2\left(\frac{3}{4}\right)^2\left(\frac{1}{4}\right)^1$ (circled)
   B $P(2) = {}_4C_3\left(\frac{2}{3}\right)^2\left(\frac{1}{3}\right)^1$
   C $P(3) = {}_3C_2\left(\frac{3}{4}\right)^1\left(\frac{1}{4}\right)^2$
   D $P(3) = {}_4C_3\left(\frac{2}{3}\right)^1\left(\frac{1}{3}\right)^2$
5. What is the probability that at least 2 of the students will buy an energy drink?
   F 42%
   G 50%
   H 75%
   (J) 84% (circled)

LESSON 11-6

Reading Strategy

Follow a Procedure

The probability that an event will happen r times can be determined by using a formula that involves the probability that the event will occur (p), the probability that it will not occur (q), a combination, and the total number of events taking place (n).

Example: In the senior class, 1 out of 4 students plays a sport. The principal chooses 2 students at random. What is the probability that both students play a sport?

Step 1	Identify n, p, q, and r.	$n = 2$; $p = \frac{1}{4}$; $q = 1 - \frac{1}{4} = \frac{3}{4}$; $r = 2$
Step 2	Write the equation and substitute for the variables.	Equation: $P(r) = {}_nC_r p^r q^{n-r}$ Substitute: $P(2) = {}_2C_2\left(\frac{1}{4}\right)^2\left(\frac{3}{4}\right)^0$
Step 3	Evaluate the combination.	${}_2C_2 = \frac{2!}{2!(2-2)!} = 1$
Step 4	Complete the solution.	$P(2) = 1\left(\frac{1}{4}\right)^2\left(\frac{3}{4}\right)^0$ $= 1\left(\frac{1}{16}\right)1 = \frac{1}{16} \approx 0.0625$

Follow the procedure to solve.

1. Of the 6 kittens in the litter, 1 in 3 has blue eyes. What is the probability that if 2 kittens are picked at random, 1 will have blue eyes?
   a. Identify n, p, q, and r. $n = 2$, $p = \frac{1}{3}$, $q = \frac{2}{3}$, $r = 1$
   b. Write the equation and substitute for the variables. $P(r) = {}_nC_r p^r q^{n-r}$; $P(1) = {}_2C_1\left(\frac{1}{3}\right)^1\left(\frac{2}{3}\right)^1$
   c. Evaluate the combination. ${}_2C_1 = 2$
   d. Complete the solution. $P(1) = \frac{4}{9} \approx 0.44$
2. Jason sees that 2 out of every 5 cars that passed his house had windshield wipers on. What is the probability that of the next 4 cars that pass his house, 2 of them will have windshield wipers on?
   a. Identify n, p, q, and r. $n = 4$, $p = \frac{2}{5}$, $q = \frac{3}{5}$, $r = 2$
   b. Write the equation and substitute for the variables. $P(r) = {}_nC_r p^r q^{n-r}$; $P(2) = {}_4C_2\left(\frac{2}{5}\right)^2\left(\frac{3}{5}\right)^2$
   c. Evaluate the combination. ${}_4C_2 = 6$
   d. Complete the solution. $P(2) = \frac{216}{625} \approx 0.35$